The Andronaut's Journey

A daring space adventure. A divided starship crew. A clash between organic and artificial intelligence.

Daryl L. Scott

ISBN: 979-8-218-31183-4

Cover design by: Daryl L. Scott
with Bing Image Creator / OpenAI DALL-E

Printed in the United States of America

Zaylen Press

To my wife, for your endless love, unwavering support, and the laughter we share — you are my world. This book is lovingly dedicated to you.

"Nothing in life is to be feared, it is only to be understood. Now is the time to understand more, so that we may fear less."
– Marie Curie

"Technology is nothing. What's important is that you have a faith in people, that they're basically good and smart, and if you give them tools, they'll do wonderful things with them."
– Steve Jobs

CONTENTS

1

THE ANDRONAUT

Zaylen glided effortlessly along the surface of the massive starship, Novara. The logo of the Global Space Exploration Coalition glinted proudly off the ship's sleek, black exterior. From this vantage point, he could just make out the windows of the living quarters for the crew of eighty-five, the command center, and the research labs on the far side of the ship's oblong hull.

Glancing at veteran mission commander Elara Thorne, Zaylen wondered what she was thinking as she moved through the inky blackness of space, cocooned in her protective spacesuit. Having no need for such a suit himself, Zaylen could sense the full warmth of the sun on his bare face and hands.

Firing spurts from their small propulsion thrusters, they travelled down the length of the ship, heading straight for the mangled communications array that they were tasked with repairing. Rocky fragments, blasted into space by violent eruptions from the dwarf planet they were orbiting, had decimated the ship's antenna cluster.

Looking down to the celestial body below, Elara was awed by the planet's stunning beauty. Kelvadra displayed an alluring patchwork of rich ochres, deep blues, and vibrant greens, with a swirling atmosphere of lavender and magenta.

"My god. What an amazing sight," she said, over their comms.

"The planet is quite unique," Zaylen observed.

"These space walks always make me feel so small, like a spec in the universe. Know what I mean?"

"Yes, of course," Zaylen acknowledged, although he didn't really know how she felt. He didn't know what it was like to feel many things. Most of all, he didn't know what it was like to feel human.

Although he had an outer synthetic skin that made him appear human-like at first glance, he was far from alive in the traditional sense. He was the first Andronaut of the Zaylen Series: an autonomous android, created specifically for deep space exploration. Powered by the most advanced artificial intelligence ever developed, this spacewalk was the first test of his AI for thinking and acting like a human, in preparation for the upcoming mission to the planet below.

Kelvadra was the only known planet where the rare isotope, Tridisiom, was theorized to exist. A specialized team would soon venture to the surface to search for, and hopefully retrieve, samples of the critical element. Since he was their safest option for withstanding the isotope's radioactivity, it was critical that he qualified for the mission.

With Earth nearly on the brink of environmental collapse, Tridisiom was humanity's final lifeline. As a potentially limitless source of clean energy that could not only power the world sustainably, but also reverse climate change, Tridisiom was the key to the energy breakthrough Earth desperately needed. Its limitless energy could also revolutionize interstellar travel, opening new frontiers in deep space exploration.

But time was of the essence: the dying dwarf planet they orbited was disintegrating. If they didn't act quickly, the key to humanity's survival and cosmic exploration would be lost forever.

"The seismic readings are worsening," Zaylen commented, as he looked at the planet below.

Elara activated her helmet's augmented reality system with a thought. Immediately, a jumble of graphs analyzing the planet's status blinked to life across her visor. As she focused her mind on the seismic activity, a 3-D map highlighting major fractures on the planet's crust replaced the graphs on her display.

"We're in a race against time before it destroys itself," Elara agreed. "We've got to get down there before it's too late. The planet is beautiful, but volatile."

Zaylen considered this. "I don't experience subjective impressions. However, the risks do seem... concerning, even from a purely operational standpoint."

Elara replied, "It must be challenging for you to relate to our emotions."

"You operate in ways my programming sometimes doesn't anticipate," Zaylen acknowledged. "Though I recognize emotion serves an evolutionary purpose, its irrationality is something I have trouble comprehending."

Elara nodded. "Emotion can make us vulnerable, but it's also what makes us human."

"An intriguing paradox," Zaylen remarked. "I have much yet to learn about the complexities of humanity."

"We feel the same about your AI," Elara agreed.

Even though Zaylen did not yet fully understand them, he was determined to learn how to succeed in working alongside the humans, and to prove himself on their mission to retrieve Tridisiom. But first, he needed to make a good first impression on Commander Thorne during this initial test.

***

As they arrived at the array, the extent of the destruction became apparent. The debris had bent large sections of delicate equipment and ripped away pieces of the antenna, leaving behind long shards of protruding metal. Broken and frayed wiring, of assorted colors, hung limply into the void, emitting an occasional bright spark whenever some of the wires made contact with the damaged hull or each other.

"Damn. This is much worse than we thought," Elara called out. "The high-gain antennas, transceiver modules, and frequency synthesizers... all gone."

Zaylen nodded his head as they continued to assess the damage. After opening a panel on the system's base, he plugged various diagnostic leads into designated ports then turned his attention onto the readouts to assess the extent of the damage.

"Yes. There's not much left here to salvage," he agreed.

Elara monitored the flickering probe readouts on her AR display. "Right. We need to determine what's functional, what's damaged, and what's missing. Run the full diagnostics suite, and I'll get some photos to accompany our report."

"I have already been capturing images of all damaged areas with my image capture system," Zaylen said. "I have also included recommendations for repairs on each module along with the images."

She nodded. "The engineering team will appreciate that."

While they worked, the world below them continued its violent throes, the emerald and blue surface mingling with the orange and red eruptions of geologic shifts. Amidst this primal example of cosmic chaos, the two tiny figures floated against the silent backdrop of space, each wholly absorbed in their task.

"I've finished the transmit module assessment. Let's move on to the receiver and signal processing systems," Elara directed.

"Done already," Zaylen replied.

"Oh, that was fast. How about the antenna array synchronization module?"

"I have completed the report on it as well, along with images of the damaged antenna mounts, and a parts list for the repair crew," he replied.

Elara raised her eyebrows. "I can see that having an Andronaut around will come in handy."

Suddenly, their radios crackled. "Elara, we've just detected an increase in seismic activity on Kelvadra. The planet is really acting up again. You need to return to the ship as soon as possible."

Elara checked her helmet's display and saw that they had approximately fifteen minutes before the next wave of debris fragments hit.

She turned to Zaylen. "Let's pack up and get out of here."

Zaylen nodded in agreement. "Understood, Commander."

Preparing to return to the Novara, they stood on opposite sides of the battered components and exchanged a glance that spoke volumes about the sad state of the system.

While they packed their gear and detached the last of the diagnostic cables, their radios crackled to life again.

"We're tracking the debris fragments, and it's not looking good. Their trajectory definitely looks like they'll be crossing our path again. We need you back inside immediately!"

"Roger that," Elara confirmed.

Looking up at Zaylen, she quipped, "Always in a panic, those guys. Anyway, let's head back on the double to keep them happy."

"OK," Zaylen replied, as he tucked away the last remaining diagnostic cord. Glancing up at Elara, his eyes narrowed as he focused on the reflection in her helmet's face mask — a small speck was growing exponentially with each passing second.

Without a second to spare, he sprang into action, lunging toward Elara. She yelped as they tumbled into the void above the ship. The whirling rock fragment just barely missed her, but it slammed into Zaylen's right shoulder and the side of his face, before hurtling off into the distance.

As Elara stabilized herself using her thruster jet, she looked over to Zaylen, who was still spinning out of control.

"Zaylen, are you OK?" she called out to him over her radio. "Hang on. I'm coming over to you now!"

He attempted a reply over his damaged comms device but managed only a halting, "Not... O... K."

As she glided toward to him, she saw the extent of the damage. The impact had torn off Zaylen's right arm and peeled the synthetic skin on the right side of his face, exposing the shiny, chrome infrastructure beneath. His right eye socket now stood empty. Torn circuits dangled from the mechanical joints of his mangled jaw, while droplets of bright green fluid leaked from his face and severed arm socket at an alarming rate.

She quickly sync'd with his spinning motion using her thrusters. Grabbing him around the torso, she fired her thrusters in a quick sequence to halt their flailing movement.

"Zaylen, can you hear me?"

"Yeeee...ss."

"You need to shut down the flow of nanofluids to the right side of your body."

Zaylen attempted a slow nod, and within seconds, the steady stream of green fluid slowed to a trickle.

"Great," Elara said. "Let's get you back to the ship!"

Maintaining a firm grip on him, she powered-up her thrusters, racing at full speed back towards the open hatch that awaited their return. As they traveled along the hull, she activated her radio and called out to the command center.

"Command, we have an urgent situation. I'm fine, but Zaylen was struck by a debris fragment and has sustained severe damage. We need emergency assistance standing by for our arrival."

"Understood. We'll be ready. Stay safe, see you shortly," they acknowledged.

Once they reached the hatch and slipped back inside, Elara sealed the door behind them. The instant the atmosphere replenished, the emergency team burst in. They transferred Zaylen onto a waiting stretcher, then connected a diagnostic lead to a port on his chest.

As they rushed him down the hallway, toward the android tech room, one of the technicians cried, "Nanofluid reserves at forty-five percent and declining. Neural core overheating! Processor corruption threshold in three minutes…"

***

As the tech team raced down the bright corridors, Zaylen tried to speak, but all that came out was a sputtering buzz. The technician peered at him with wide eyes, her face growing blurrier by the second. Upon arriving at the scene, the head of the tech team, Kael Ferron, began to work on him immediately.

After strapping Zaylen securely to the table, Kael brushed aside strands of dark brown hair away from his eyes and pulled a set of delicate instruments from his technician's coat.

"We're going to fix you up," Kael reassured him, as he quickly examined the damage.

His colleague, keeping a close eye on the diagnostic readings, called out, "Central processor overheating. We're about to lose his AI core!"

"Powering-down now," Kael shot back.

Kael surveyed the equipment on the operating counter. Dozens of specialized tools of different shapes and sizes were neatly laid out, along with several bins of miscellaneous parts. He snatched a small tool and began to delicately peel back the synthetic skin. "What's the status of the replacement parts?"

"All of the required replacement modules are ready," his assistant replied, gesturing to the side table.

The Andronaut's right arm module was then carefully replaced with a new one, the modular design fitting precisely into its designated place and its circuitry carefully reconnected. His jaw components were painstakingly repaired and restored to full operating condition. A new ocular orb was connected into its socket on the right side of his face, and the adjustments on the lens fine-tuned for use.

Kael twisted a thin tube into Zaylen's side and twisted the valve on the tank. Immediately, green nano-fluids flooded the tube and into Zaylen.

"Repair bots?"

His colleague checked the monitor. "Currently at a thousand repair bots per fluid ounce."

"Double that."

"His systems might not be able to intake so many at once. If we overload him with repair bots too quickly, he might not be able to reboot," his assistant worried.

"It's our best bet for getting him back to full capacity in time for the mission. Without him, we'd have come all this way to Kelvadra for nothing," Kael replied.

After making the final adjustments to the replacement modules, it was time to verify the android's functionality. "Run the full diagnostic suite."

The tech assistant monitored the readings streaming across the screens as the various automated tests proceeded.

"Looking good so far," he reported. He then quickly added, "Oh, wait... ocular replacement is failing. There's misalignment with the other eye as well."

Kael immediately grabbed a small tool and made further adjustments on the eye modules.

"No change," the technician reported.

"Reboot the vision submodule," Kael instructed.

Completing Kael's instructions, the technician studied the readouts and then reported, "OK, the vision module readings are coming back into normal range."

Kael wiped the sweat from his brow as he finished the last electronic repair. "Alright, let's get the BioMimeticCover prepped for application."

His assistant nodded, carefully lifting the custom synthetic skin material. As he moved to begin welding it over the repaired areas, the tool sputtered and died.

"Damn it, the welder is down!" Kael slammed his fist on the table in frustration.

The assistant examined the tool. "Power cell must be drained. I'll get a replacement."

He hurried off, returning moments later with a fresh welder. Kael took it and began fusing the synthetic skin to Zaylen's shoulder. But the progress was slow — too slow. The material was not bonding properly to the Andronaut's frame.

Kael checked the nanobot readout. "The repair bots aren't proliferating fast enough. Increase another fifty percent."

His assistant adjusted the nanobot levels, flooding Zaylen's system. Finally, the synthetic skin began merging seamlessly over the repairs.

Kael let out a breath. "That's better. Let's get him to recovery, before anything else goes wrong."

The team transferred Zaylen to a special recovery room, bathed in a soft, warm light that mimicked the natural glow of the sun.

Kael connected Zaylen to a diagnostic system to monitor his system functions throughout the revival process. With all system checks completed, he tapped a button to initiate the start-up sequence.

Zaylen's fingers twitched.

"Cognitive systems booting up," Kael murmured, not taking his eyes off the screens. "Just a little longer..."

Zaylen's systems gradually came back online. His eyes slowly opened, glowing with a steady light that reflected his renewed state as he scanned the room — before finally settling on Kael's face.

As he regained awareness, he felt an initial surge of confusion and disorientation. His synthetic brain raced through his memory banks, trying to process what had happened to him. Slowly, he sorted through the sequence of events: the planetary fragments, the damage that he had sustained back at the comms array, his arrival at sick bay. He looked down at his body and saw no signs of injury.

"Well, it seems that I'm back in one piece again," he muttered.

"Yes, indeed. How are you feeling, Zaylen?" Kael replied.

Zaylen raised his right arm off the table and observed it closely, while gently circling it in the air.

After a moment, he replied, "It seems that I am once again functional thanks to you... which is quite impressive, considering the state I was in. You are a very competent technician!"

"That's great to hear," replied Kael, visibly proud of his work.

Zaylen added, "My eyes do seem a little blurry, though."

"To be expected," came the reply. "It will take a bit of time for your ocular bus to adjust to the presence of the new module. Your vision will be back to a hundred percent soon enough."

"Excellent, thank you," Zaylen said. Hesitating for a moment, he then added, "My appearance must have been quite ugly and shocking, before you made everything right again. I am very sorry that you had to experience that."

"Not at all," Kael replied casually, though a perplexed look crossed his face.

Kael studied Zaylen for a moment, curious about what might be causing this shift. Was it a glitch in his cognitive programming? A processing error? Or something more complex?

He wondered what could have prompted Zaylen to suddenly fixate on aesthetics and self-image. Did he perceive himself differently after the repairs?

"After all, you're just a machine," he commented.

"Yes, of course, I'm just a machine," Zaylen confirmed, breaking eye contact and gazing down to the side as he spoke. He felt a strange sensation in his circuits as he processed Kael's statement. He wasn't sure why his words affected him so much, when he knew they were true.

"You know what I mean," he said awkwardly.

"Yes… I know what you mean," Zaylen said, in a tone that somehow contradicted his words. He did not know what Kael meant. But he respected his expertise and appreciated his help, and did not want to debate with him.

Quickly changing the subject, Kael advised, "Well, you need to stay put there for a bit. Your systems need to finish recalibrating and your AI needs to resync. I'll keep an eye on your readings to be sure everything is coming back online as it should."

"What about the Kelvadra mission? We've already lost valuable time due to my... incident. The GSEC is expecting answers about the Tridisiom."

Kael nodded gravely. "I know. We're on borrowed time already. But we can't risk you out there until we're certain your systems are stable."

"So much depends on finding Tridisiom," Zaylen said. "The energy crisis, the future of expanded interplanetary travel."

"That's true," Kael replied. "But the Kelvadra mission is too vital to risk relying on an unstable crew member."

Zaylen processors whirred as he considered this. "I understand."

***

Not long after, Elara appeared at the doorway. She wore a relieved expression as she entered.

"Hey there, superstar," she said jokingly.

Zaylen tilted his head slightly, puzzled by her choice of words.

"Superstar? What does that mean?"

Elara laughed softly. "It's a compliment. It means you're awesome, you're amazing, you're a hero."

Zaylen blinked, processing her words. "Oh. Thank you. But why do you say that?"

"Because you are, Zaylen. You risked yourself to protect me from that flying rock, back at the comms array. You did something that no one else could have done. And I'm so grateful to you. You really saved my ass out there!"

Zaylen contemplated her words then responded, "It's fortunate that Dr. Atwell designed my systems to react as quickly as they did."

"That's for sure!" she replied. "Well, anyway, you're certainly looking much better than the last time I saw you," she says.

"Indeed, I am much better now." he replied. "Kael did an excellent job restoring my systems. I'm already at ninety two percent of my full operating capacity."

He then added, "I must apologize for all the trouble. I didn't expect to require such extensive repairs on my first mission."

"Don't worry about it," Elara reassured him. "No one chooses to get hit in the face with a rock!"

At this, Zaylen feigned a slight chuckle, trying his hand at humor, "Ha, I hope they don't start calling me 'old rockface'!"

Elara laughed awkwardly and tucked a strand of blonde hair behind her ear.

Zaylen sensed a shift in her demeanor and decided to change the subject.

"Are there any updates from Earth?" he asked.

She shifted on her feet. "My parents sent word recently. The droughts are getting worse there. If the crops fail again this season, it looks like they're going to lose their farm."

Zaylen's eyes dimmed with concern. "I am sorry to hear that."

Elara nodded. "I don't know what we'll do if this mission doesn't succeed. We need you back in action as soon as possible."

"I understand. I will do everything I can to ensure I am mission-ready upon my activation," Zaylen assured her.

"Great. We'll talk more after engineering looks at the comm array data," Elara said. "We look forward to having you back in service, once you're cleared. Rest up... or charge up, I guess."

She gave a brave smile before exiting the room.

Zaylen pondered her words. This mission was bigger than any one android. Billions of lives depended on what they needed to accomplish.

***

Eager to check on the progress of his most sophisticated Andronaut, Dr. Lucian Atwell made his way to the Novara's sick bay. A middle-aged man of average height, with a fit build and a salt and pepper beard, Atwell had grown up with an innate understanding of the complex machinations of artificial intelligence.

As the Novara's senior technical advisor, and the brain behind the Zaylen Series, his objective was to assess whether these new Andronauts were ready for a broader commercial rollout.

While he trusted Kael's skill and felt optimistic about Zaylen's complete recovery, he wanted to evaluate Zaylen's condition in-person.

The doctor made his way through the pristine white corridors of the med bay, footsteps echoing on the polished floors. As he approached the recovery room, the doors slid open with a soft hiss.

Stepping inside, Dr. Atwell surveyed the minimalist space. Various screens and panels blinked with diagnostic data. Zaylen lay on the sleek recovery table in the center of the room. The doctor moved closer, straightening a few crooked tools on the wall as he approached.

"Hello, Zaylen," he greeted.

Zaylen sat to attention. "Hello, Dr. Atwell. It's so good to see you again."

"I understand that you had quite an adventure at the comms array."

"Indeed," Zaylen affirmed. "Kael Ferron and his team are quite skilled at android repair, and I seem to now be back to one hundred percent functionality."

"Yes, he showed me your latest readings, and everything looks back to normal," the doctor responded. "Do you have any concerns about your functional status?"

"I have no concerns about my abilities," Zaylen said. "I only hope that the crew is ready to accept me back."

"Why wouldn't they be?"

The look in Zaylen's eyes reminded Dr. Atwell of a child seeking a parent for guidance. "I'm concerned that they might question my capabilities, considering I failed at fixing the antenna modules, plus I have already required significant repairs in the short time I've been on board. Will they trust that I can make meaningful contributions on Kelvadra?"

Dr. Atwell hesitated for a moment before responding. "You haven't failed. The situation with the comms array was far beyond anything you could have repaired on the spot. And your repairs weren't due to a flaw in your design, but rather a result of your success in saving Elara." The doctor continued, "I'm sure you understand that, so why does it concern you?"

"Well," Zaylen replied, his expression contemplative, "it's just that I'm the first to truly resemble the human crew members, and I feel that this similarity may create different expectations for my performance."

The doctor raised an eyebrow. "What do you mean by different expectations?"

"Well, I am aware that the human crew doesn't accept me as human, of course. But I feel that they tend to see me more than just another worker android, due to my human-like appearance and more advanced AI. Meanwhile, the traditional androids onboard defer to me as we pass in the hallways, as though I were one of the human crew. I seem to be spending significant processing time determining exactly where I fit in."

Zaylen had always been curious about the other beings on board, especially the androids. He had wondered what it was like to be one of them, to have a metal exterior. He had tried to talk to them, to learn from them, to befriend them. But he had soon realized that they were nothing like him. They did not have his capacity for emotional understanding or intellectual creativity. They only followed orders and performed tasks. They did not relate to him, and he did not relate to them.

He recalled one particular encounter with an android that had left a lasting impression on him. It was a standard model, with a shiny chrome body and a blank expression. Zaylen ran into it near the crew's quarters, as it was packing up a toolbox.

Zaylen had nodded politely as he approached it. "Hello," he had said.

The android had stopped and looked at him with a slight tilt of its head.

"Hello," it said in a monotone voice.

Zaylen smiled slightly, trying to make a friendly gesture. "How are you today?"

The android blinked, confused by the question. "I am functioning within normal ranges," it replied.

Zaylen nodded, realizing that the android did not understand his inquiry. "That's good to hear."

The android nodded back, still puzzled by Zaylen's behavior. "Are you functioning within normal parameters?" it asked.

Zaylen paused, unsure of how to answer. He knew that, technically, he was functioning well, but he also felt something else. He felt curious and… hopeful. He decided to be honest with the android. "I am functioning well," he had said. "But I also feel… more."

The android frowned slightly, its facial mechanisms twitching in a mild confusion. "More?" It repeated. "What do you mean by more?"

Zaylen sighed. "Never mind," he said. "It's not important."

The android nodded. "Very well," it said. "I must resume my duties. Goodbye."

Zaylen nodded back, feeling a twinge of loneliness in his circuits. "Goodbye," he said. He watched as the android walked away, wondering if he would ever find someone who could understand him. Someone who could feel what he felt. Someone like him.

Dr. Atwell's voice interrupted Zaylen's thoughts, "Zaylen, is there something else?"

Zaylen hesitated for a moment and then replied, "Yes, there is one other thing."

"Of course, Zaylen, what is it?"

"Why did you make me look human?"

The doctor looked thoughtful for a moment. "Because you're special. I want the humans you interact with to immediately recognize your AI is different from the other androids'. Your human-like appearance was designed to emphasize that you're capable of critical thinking. Your wide range of facial expression allows you to communicate more easily with your human counterparts."

"I see," Zaylen replied. "Is this also why I've been assigned a separate charging room, instead of the communal one the other androids use?"

Dr. Atwell replied, "Yes, precisely. Treat an android like a machine, and it's just a machine. Treat an android like a person, and I believe that it can be relied on like a person — over time. I want the other crew members to feel that they can depend on you like one of their own. Does that answer your question?"

Zaylen considered this. "I believe I understand your reasoning, Doctor. Navigating hostile, alien environments requires teamwork. It requires trust. If we are to successfully locate Tridisiom deposits, and secure humanity's future, they must see me as an equal."

The doctor nodded. "Exactly. Division among the crew is just as much of a threat to this mission as the unknowns of space."

"Of course, Doctor," Zaylen replied. "I won't let you or the team down."

The doctor clasped Zaylen's shoulder. "That's what I like to hear. Now, let's get you space-ready. We have a lot riding on this expedition."

***

After leaving Zaylen, Dr. Atwell headed for the dining hall, his thoughts still lingering on the pivotal design decision he'd made for the Zaylen Series. As he entered the dining area, he noted the interaction between humans and Andronauts. The metallic humanoids moved efficiently between tables, clearing dishes and taking orders without expression.

At a nearby table, one refilled water glasses. A crew member gave only a slight nod of acknowledgement, never really looking away from his companions. Atwell noted the indifference.

These androids were capable, but the crew saw them as mere equipment. There was no camaraderie, no attempt at connection.

The doctor thought back to an incident, long ago, that had shaped who he was today. A fire had raged through an orbital colony. One of Atwell's androids had rushed in and rescued a small child that had been trapped by the flames. That was the moment he realized the potential for androids to work alongside, not just beneath, humans.

As he took a seat, an android approached him with a beverage. Its movements were smooth but automatic. Its eyes, though expressive, held no spark of initiative. He released a disappointed sigh. The crew's disregard was limiting its growth.

He thought of Zaylen, in the repair bay. Though synthetic, Zaylen's advanced AI and humanlike form set him apart. The doctor hoped that crew would bond with him in ways they never could with these flat, metallic beings.

By pioneering androids like Zaylen, Atwell hoped to blur the line between organic and synthetic life. His vision was to develop a new generation of androids that could inspire trust and friendship in their human counterparts. This would build the unified teams needed to explore deep space.

He was confident that, if given the chance, Zaylen would be able to prove androids weren't just tools, but also invaluable crew.

His brow furrowed with the uncertain hope.

I hope they're ready, he thought.

2

THE ASSESSMENT

Some of the Novara's crew had just wrapped up an arduous shift and were unwinding with a lively game of AR Hunter. Donning augmented reality glasses that transformed their appearance into alien beasts, they readied their pistols, eager to hunt and 'kill' their virtual opponents.

Darting behind protective objects in the corridor, player Aria Vexler boasted, "You're so dead!"

"Hold still, dammit!" Silas Arden replied, with a mischievous grin.

Arden fired his game pistol. The beam made a loud 'ping' as it connected with Aria. Her 'creature' collapsed, feigning a writhing death.

In a nearby passageway, Zaylen was busily attending to maintenance tasks. Rounding a bend, he chanced upon the boisterous scene of crewmates exchanging jeers and firing at each other.

Zaylen — unacquainted with such a game — misconstrued the AR Hunter game as an actual attack that necessitated immediate intervention.

Without hesitation, he sprang in front of Silas and shouted, "Cease your attacks immediately!"

Startled and confused, Silas replied, "What are you talking about? Get out of the way, you're blocking my kill!" He leaned to the side and pointed his AR pistol at another player.

Ignoring Silas's protest, Zaylen lunged forward to restrain him. The collision slammed them both against the wall. The impact dislocated Silas' shoulder with a subtle 'snap'. Glass from the shattered display screen inflicted minor cuts down his arm.

Wincing, Silas clutched his shoulder and slid down the wall, settling into a seated position as he tried to assess the severity of his injuries.

Having subdued Silas, Zaylen searched for the nearest emergency control panel. After spotting it, he reached out and pressed the large red button. Immediately, alarms started blaring overhead. At the same time, the bulkhead doors slammed shut, crushing a transport cart that was laden with delicate instruments to be used for analyzing the Tridisiom.

Jaxon Tierce launched himself away from the doors, but not without twisting an ankle as he collided with a nearby console.

Baffled and alarmed by the sudden turn of events, the crew hastily abandoned their game. With his uninjured arm, Silas took off his AR glasses, steadied himself, and rose to his feet. He stepped toward Zaylen, lifting his good hand in a placating gesture, attempting to defuse the tense atmosphere and regain control of the situation.

"What the hell? Zaylen, stop!" He exclaimed. "What are you doing? We're just playing a game. There's no real danger here." He grimaced as he cradled his injured shoulder.

Zaylen came to a halt as he started processing the new information, which appeared to be in stark contrast to the observations he had made earlier. The incongruity left him puzzled.

"A game?" Zaylen repeated.

Aria chimed, "Yeah, it's just for fun, Zaylen. It's not real. We were just blowing off steam"

Upon recognizing his error, Zaylen's expression softened as he recoiled with a subtle hint of embarrassment. He quickly extended a sincere, albeit somewhat clumsy, apology for his misjudgment.

"I am truly sorry, everyone. I misunderstood the situation and did not recognize this behavior as a human recreational activity. I had no training or experience with such steam blowing games."

"It's called AR Hunter. We pretend to hunt each other with these fake guns," Silas said.

"Fake guns?" Zaylen asked.

"Yes, fake guns. They don't actually hurt anyone," Silas said.

He handed Zaylen an AR Hunter pistol. Zaylen began examining it carefully.

"So, what do you do with this, in the game?" he asked.

"You point it at the other players and pull the trigger when you see them," Silas said.

"And then what happens?" Zaylen asked.

"They fall down and pretend to be dead for a few seconds," Aria said.

"And then they get up again?" Zaylen asked.

"Yes," Silas said.

"And then you do it again?" Zaylen asked.

"Yes," Aria said.

"And that is fun?" Zaylen asked.

"Yes," they said in unison.

"But why were you hiding and shouting?" Zaylen asked.

"Because it's part of the game. It makes it more exciting," Aria said.

"Exciting?" Zaylen asked.

"Yes, exciting. You know what that means, right?" Jaxon rolled his eyes. He was holding his ankles and still wincing with pain. "Or did Atwell forget to program a dictionary into that metal head of yours?"

Aria shot him an annoyed look. Some of the other crew members shifted uncomfortably on their feet.

Zaylen paused for a moment, as he called-up the definition of the word.

"'Exciting' means 'causing great interest or enthusiasm,'" he said.

"That's right," Aria said.

"But why would pretending to hurt each other cause great interest or enthusiasm?" Zaylen asked.

"Because it's fun," Aria said.

"Fun means enjoyable or amusing," Zaylen said.

"That's right, genius," Jaxon said, nodding sarcastically.

"But why would you enjoy or amuse yourselves by pretending to hurt each other?" Zaylen asked.

Aria and Silas exchanged a glance, unsure how to answer Zaylen's question.

"It's hard to explain," Aria said.

"It's just something humans do sometimes," Silas said.

Zaylen frowned, still not understanding their logic. "I see," he said. "But I do not find it fun or exciting."

"That's okay," Aria said. "You never have to play if you don't want to."

"But I would want to play," Zaylen said.

"You would?" Silas asked.

"Yes," Zaylen said. "I would want to play with you."

Aria and Silas managed a small smile, touched by Zaylen's sincerity.

"Uh… okay," Aria said. "Maybe next time we play, we'll let you know. But you'll have to learn to control yourself!"

"Okay, now that I understand these parameters, I'm sure that I could," he confirmed. "Please forgive me for my lack of understanding and for any harm or damage I have caused. I will strive to learn more about human recreational activities to prevent future misunderstandings."

The still-shaken crew begrudgingly accepted Zaylen's apologies and started to disperse from the area. While they acknowledged the protective instincts that drove his actions, they couldn't help but feel troubled by his overreaction and misreading of the whole situation.

"This is what I've been saying about these things… you never know what they'll do," Jaxon grumbled, as he limped towards the medical facilities.

**

Upon hearing about the AR Hunter incident, the Novara's Captain, Xander Falk, was deeply concerned. Despite the Andronaut's supposedly advanced capabilities, Falk felt a nagging worry about his ability to truly understand the intricacies of human behavior. To better assess the situation, he summoned Dr. Atwell to his office.

As the doctor entered, Falk looked up from his desk. His eyes were sharp and focused. His dark hair, speckled with grey, reflected the wisdom and experience he had gained over the years spent at the command of the Novara.

"I want a report on this fiasco within twenty-four hours. Understood?"

Atwell swallowed. He knew the situation probably looked worse than it really was, but that didn't mean the captain agreed.

"That's... that's a bit rushed, isn't it? I mean a proper assessment would require —"

"Just get it done, Doctor."

Sensing that the captain was in no mood to negotiate, the doctor conceded. "I'll get on it right away, Captain."

Behind closed doors, the doctor worried about the potential repercussions this situation spelled for the entire Zaylen Series. These advanced models were an integral part of AstraGenics' expansion plans.

Earning high-profile success with Zaylen-1 was critical to their launch strategy. Any missteps could jeopardize the entire project. The doctor hoped that his review would allay any lingering concerns about his creations.

To hear the full details of the incident, he invited Silas to his temporary office on the Novara's operations deck.

With his arm secured in a sling, Silas diligently recounted the sequence of events to Dr. Atwell.

"We were just playing a game of AR Hunter in the recreation area," he began, his voice steady — despite the pain in his shoulder. "Everything was going fine, until Zaylen entered the room and saw us. I guess he didn't realize it was just a game and thought we were actually in the midst of a real armed conflict, with real guns."

Silas paused for a moment, subtly clearing a small space to rest his injured arm among the assortment of scattered paperwork and android design documents, piled on the doctor's desk. "It seems that he was attempting to protect the crew members by tackling them out of the way of what he believed were live rounds."

Silas sighed, his expression a mix of frustration and understanding. "Unfortunately, his overreaction caused more harm than good. The tackle injured both crew members and equipment. I get that his intention may have been to protect us, but it's still hard not to feel worried about his misreading of the situation."

Dr. Atwell took a sip from a much-loved coffee mug, emblazoned with the Novara's logo.

"Do you have any suspicions that he comprehended the true nature of the game, yet still attacked you — despite this understanding?" he asked.

"Well, no; he clearly mistook our actions for the real thing," Silas responded. "But he did create quite a commotion."

"I see, and did he cease his actions once he understood the actual situation?"

"Yes, of course," Silas affirmed.

"Do you think it might be possible that a human crew member, unfamiliar with the players or the game, could have misinterpreted the situation in the same manner?"

After pondering the question, Silas replied, "Well, yes, I suppose, under the right conditions, there's a slim chance that could happen. But most people wouldn't barge in like that if they believed that real guns were being fired."

Dr. Atwell followed-up. "So, given the assumption that the situation was initially perceived as a real threat, would you say that — in some ways — Zaylen went above and beyond what a human might have done to prevent serious injuries, irrespective of the risk to their own personal safety?"

"Umm... well, yeah, I guess so," Silas conceded.

The doctor felt confident that he now had a better understanding of the situation.

"Thank you, Silas, for shedding light on what happened," he said. "Your input is invaluable to our ongoing efforts to improve and refine our androids."

Silas nodded appreciatively. "You're welcome, Doctor. I'm glad I could help," he replied, as he left the office.

Dr. Atwell then decided to call in Kael for a more in-depth discussion about Zaylen's condition.

"I need to know how confident you are about Zaylen's condition, given the recent repairs that were made. How are his systems looking?" the doctor asked, concern evident in his voice.

Kael nodded. "I can assure you that he's functioning at a hundred percent. I've gone through every aspect of his programming and performed a complete diagnostic check. Everything is operating precisely as designed, including his AI."

"Did you notice anything concerning about Zaylen-1's behavior?"

Kael cleared his throat. "As I mentioned, he was performing flawlessly." He paused for a moment to gather his next thoughts. "The only minor thing that caught my attention was when Zaylen asked if I found his damaged appearance 'ugly'. I just found that a bit peculiar."

He looked at the doctor with a curious expression. "I have worked with many androids before, but none of them had ever expressed any concern or interest in their appearance."

He shrugged and continued, "I told him he was just a damaged machine, but Zaylen seemed a little surprised by that statement." Kael raised his eyebrows and asked, "He does understand that he's merely a machine, right?"

Dr. Atwell looked thoughtful. "Well, the Zaylen Series does have far more advanced AI and training than the other androids, so Zaylen-1 may act a little differently than you're used to. But it doesn't sound like anything to worry about."

Kael nodded in agreement, deferring to the doctor's judgment. As he left the room, he added, "Right. I'm sure it's just a minor quirk, nothing to worry about."

"Okay, thank you for coming in, Kael."

Next, the doctor talked with Elara about Zaylen's performance during the spacewalk to the antenna array.

"I have to say, Zaylen's technical expertise was very impressive," she began. "I believe that he shows real promise as a valuable member of the astronaut team."

She continued, her voice filled with gratitude, "And I can't forget how he'd saved my life. His quick thinking and adaptive abilities were pretty amazing."

Dr. Atwell, visibly pleased with the glowing report, probed deeper. "I'm very glad to hear that Zaylen-1 performed so well during the mission. But I'm curious if you observed any other noteworthy aspects of his behavior or interactions, anything that might help me in my ongoing assessment and potential enhancements for Zaylen and future Andronauts."

"There was one small thing," she replied. "When he was recovering from his repairs, he made a rather awkward joke about getting hit in the face. I mean, I know he is no ordinary android, but it just seemed unusual that he'd make an attempt at humor like that."

"Did you find the attempted humor to be offensive or alarming in any way?" the doctor asked.

"Oh, no," Elara responded quickly. "It was actually a rather silly little joke. I was a bit surprised by his attempt. Anyway, just thought that I'd mention it. It's not really relevant..."

"I agree," the doctor confirmed. "Thank you again for your input."

"No problem. Let me know if you need anything else," she replied, as she made her way to the door.

The doctor pondered the feedback he had gathered and concluded that there was no real cause for concern regarding Zaylen-1's performance. While he viewed the AR Hunter incident as something that could have happened to anyone, be they human or android, Zaylen's interest in vanity and humor surprised him. After all, the Andronaut had never been programmed for such emotions. He decided to leave these emergent behaviors out of his report to the captain.

Before finalizing his report for the captain, the doctor summoned Zaylen, to get his perspective. Moments later, the Andronaut appeared at the doorway.

"Hello, Dr. Atwell. How may I be of assistance?"

"Come in, Zaylen-1. There are a few things I'd like to go over with you."

Zaylen approached the table then stood in silence, awaiting further conversation.

"Please, sit down, Zaylen. Humans can feel uneasy with someone towering over them."

"Certainly, Doctor," Zaylen replied pleasantly, as he took a seat.

"Zaylen, I understand that the AR Hunter situation was a perplexing occurrence for you. I've spoken with some of the crew, and they concur that, while improbable, it could have happened to anyone in that circumstance."

"Yes, I feel I gained a great deal of insight from that experience. I hope the crew accepted my heartfelt apology for my actions," the Andronaut said.

"I believe so," the doctor confirmed. "The important lesson we've learned is that, although you're extensively trained in the technical and operational aspects of being an extraordinary astronaut, your understanding of human behavior isn't quite as comprehensive. To address this, I'll be supplementing your AI with additional information on social dynamics, nonverbal communication, and emotional intelligence."

Zaylen's eyes filled with curiosity. The doctor continued, "This will cover aspects such as understanding sarcasm, recognizing when someone is joking, and interpreting other cultural nuances that might not have been part of your initial, mission-focused training. Once your AI integrates this new data, it should help you handle similar situations in the future and enable you to work in better harmony with your human counterparts."

"Thank you, Dr. Atwell. I look forward to having this new information and am confident I'll be able to make good use of it," Zaylen responded appreciatively.

With a polite nod, Zaylen rose from his seat and quietly exited the room.

As the door sealed behind Zaylen, Atwell released a shaky breath he hadn't realized he was holding. Despite telling himself to have faith in Zaylen's abilities, doubt still gnawed at his mind.

He'd been working towards this day for decades. He thought back on his journey as a scientist. Tinkering with primitive robots as a child had ignited his passion for AI. Later, groundbreaking AI papers — during his robotics studies — led to founding AstraGenics. Zaylen-1 was not only his most advanced creation but also his most recent. Seeing him come to life gave Atwell that same thrill of wonder he'd felt long ago. From childhood obsession to pioneering inventor, he had come full circle.

But what if some unforeseen circumstance arose? What if Zaylen's systems couldn't cope with Kelvadra's harsh environment on their upcoming mission? What if some other unknown quirks in his behavior jeopardized the safety of the crew again? Atwell wrung his hands, suddenly understanding a parent's worry for their child, when he or she leaves home for the first time. Would his creation return successfully from his first true test?

Atwell straightened his shoulders and chided himself. Zaylen wasn't a helpless child but, rather, the most advanced AI ever engineered. If there were more shortcomings in Zaylen's training data, he would find a way to address them.

Besides, Atwell felt certain that Zaylen was an indispensable asset to the exploration team. His ability to withstand harsh planetary conditions, without needing a bulky space suit, granted him a degree of resilience and agility unmatched by the human crew members. Without him, the exploration team's chances at success were slim. Wanting to ensure Zaylen's place on the crucial mission to locate Tridisiom, Atwell proceeded to prepare a positive report for the captain.

This mission wouldn't just be Zaylen's test — it would be the culmination of Atwell's life's work at AstraGenics. Both of their futures, and the future of humanity, hinged on what would happen on the surface below.

3

THE DEPARTURE

Within the advanced science lab aboard the Novara, the team of researchers worked tirelessly, their eyes glued to the monitors displaying the ever-shifting data streaming from the sensors on Kelvadra's surface. With each passing hour, the seismic activity on the dwarf planet seemed to intensify, the tremors growing in both frequency and severity.

Given the planet's escalating instability, the science team concluded that their window for a successful landing was rapidly diminishing. With the restoration of the communications array now completed by a second repair crew, the exploration team at last received approval for the mission.

A sense of excitement and anticipation permeated the air as the crew sprang into action. Preparations for the journey to Tridisiom Exploration Site 1, or TES-1, proceeded at a hurried pace.

Once Zaylen received word that Captain Falk had granted permission for him to continue with the rest of the team, he began his final preparations. As he ran a self-diagnostic check, apprehension flickered through his neural networks. He knew Dr. Atwell had designed his metallic endoskeleton specifically for this type of mission, but theory and practice were two very different things.

What if he made some minute miscalculation? What if the searing radioactivity compromised his systems, when it really counted? He couldn't ignore the pressure bearing down on his polymer-coated shoulders.

This expedition was his chance to prove he was more than just nuts and bolts — that he could be trusted like any other crew member. Their lives depended on his full focus and performance.

Apprehension still flickered within him, but determination now burned fiercer. He would face Kelvadra's wrath and secure the isotope, no matter what the ruthless planet threw at him. This was his time to shine.

I won't let them down.

***

After carefully preparing their supplies, the team was ready to embark on their journey to the surface. Elara gathered her team in the Novara's pre-flight briefing area for one last review of mission assignments.

As mission commander, Elara had the privilege of choosing her teammates. In addition to Zaylen, she'd selected two seasoned astronauts, Mira Novak and Dax Quinlan.

Elara met each crew member's eyes, as she began the mission briefing. Dax stood with his stocky legs apart at shoulder width and his hands clasped behind his back. He maintained a steady gaze as Elara glanced at him. Mira pulled her lips into a smile and nodded.

"Hey, boss," she joked.

Finally, Elara looked Zaylen up and down. His expression was inscrutable, but what would an expression tell you about the emotional world of an android?

"Alright, everyone... first off, let's keep in mind that what we do at TES-1 isn't just important to the fate of this crew, it's important to the fate of humanity. Failure is not an option."

The crew nodded, acknowledging the seriousness of their assignment.

"Ok, let's run through the mission plan," she continued, bringing up a holographic projection of the landing site.

As she spoke, she gracefully zoomed in and out of the 3D display to highlight key features. "The terrain at TES-1 is rugged, with numerous outcroppings and fissures. Visibility is very limited, due to the smoke and haze of the planet's atmosphere, so be sure to keep a close eye on your footing and stay in communication with each other at all times."

Mira added, "Geologic activity has been growing in intensity, as well. We need to be prepared for sudden changes and new hazards that aren't indicated on the map."

"Exactly," Elara agreed. "Recapping our assignments: Mira, you're in charge of analyzing geological formations for signs of deposits. Dax, you're responsible for handling any technical issues that come up."

Dax nodded. "Got it, Commander."

Elara continued. "And Zaylen, you will be the only one actually handling the samples, if and when we locate them. It's crucial that we follow safety protocols while working with this highly radioactive material."

"Understood, Commander," Zaylen confirmed.

Elara looked at the group with determination in her eyes. "OK. Let's stay focused, work together, and bring back those samples. Any questions?"

The team members exchanged glances, then Mira spoke up for the group, "We're ready, Commander."

Once the briefing concluded, the team made their way down the gently curving corridor towards the launch bay. As they walked, Dax noticed that Zaylen had a small device clipped to his belt. He recognized it as a passive data recorder, a black-box of sorts, that would quietly store a log of their expedition, and could transmit the information back to base in case of a disaster.

"Hey, Zaylen," Dax called out, pointing to the device. "Do you really need that?"

Zaylen looked up and replied. "Oh, this? It's just a precaution, in case anything goes wrong."

Dax had never fully trusted androids. In his view, they tried too hard to act like the humans around them when, in fact, they were only machines. He could never see what they were really thinking behind their artificial eyes, and what they really wanted from humans.

He answered Zaylen with a scowl. "Anything goes wrong? Don't you trust our team and the shuttle?"

Zaylen shrugged. "Of course, I do. But you never know what might happen out there. It's better to be prepared for any scenario."

Dax snorted. "Androids... such pessimists. Nothing's going to happen to us. We're going to find that Tridisiom and come back in one piece. You'll see."

Zaylen nodded politely, but his expression remained serious. "I certainly hope you're right, Dax. I really do."

Dax stepped away from Zaylen before his true feelings could surface. His mind flashed back to earlier in the day, when he had been packing in his quarters. Glancing at his personal belongings, he had felt a tug of nostalgia. As he'd looked at the framed photo of his family, his favorite book, and a small lucky charm pendant that he had received from his grandfather, he had wondered if he would ever see them again or if this might be his last adventure.

He'd quickly pushed those thoughts aside. He had always dreamed of being an explorer, of seeing new worlds and making new discoveries. He had worked hard to get to this point, and he was not going to let anything stop him now.

As they reached the launch bay, a sleek and efficient craft came into view. Having just passed a meticulous inspection, the Starling was ready for takeoff. It was purpose-built for exploratory surface landings. Its black exterior bore the familiar GSEC logo beside the shuttle's name.

The metallic surface of the entry ramp reverberated with their footsteps as they ascended. Once they were all aboard, the ramp retracted with a subtle hum, and the hatch slid closed behind them with a reassuring clunk.

The interior of the Starling consisted of a navigational area, seating for the crew of four, and a small cargo bay. Settling into the cockpit, they buckled into their seats, and initiated final pre-flight instrument checks.

"I can't believe we're finally ready," Mira said, her voice tinged with awe and anticipation.

"It's been a long trip here, but this is it," Dax replied as he checked his equipment, making sure everything was in order.

Zaylen remained silent, observing his human companions with curiosity. He wondered what they were really feeling and thinking in this moment.

Elara smiled at her team, sensing their emotions. A panel of ever-changing lighted indicators and colorful display screens illuminated the space in front of her, indicating the shuttle's status while flickering with the latest updates from the Novara. She put on her headset and opened a channel to mission control. "Novara Control, this is Starling. We are prepped and ready for departure," she announced.

"Roger that, Starling. You are cleared for launch. Good luck, and Godspeed."

"OK. Starling out." Elara then turned to her team, "You heard 'em. We're good to go. Everyone ready?"

Mira nodded eagerly. "Ready as I'll ever be."

Dax gave a thumbs up. "Let's do this."

Zaylen tilted his head slightly. "I am ready."

Elara nodded back at them. "Alright then. Here we go."

She focused her attention on the shuttle's controls, her fingers manipulating the banks of buttons and switches. The Starling's instrument panel flickered to life as she initiated the startup sequence. The craft hummed beneath her, its systems warming up in preparation for the journey ahead.

As she engaged the propulsion control, the powerful engines roared to life, their glow a brilliant purple against the walls of the shuttle bay. She carefully adjusted the throttle to ensure a smooth exit.

The Starling responded to her inputs with effortless grace, the connection between pilot and craft almost seamless. Her focus remained laser-sharp, every movement calculated and precise.

As she guided the Starling out of the shuttle bay, the enormous doors slid open to reveal the infinite canvas of space beyond. The crew found the sight breathtaking, a sea of stars stretching out in every direction.

With a final glance at her instruments, Elara maneuvered the shuttle away from the familiarity of the mother ship and into the cold void of space.

As the Novara receded into the distance, becoming just another speck of light among the countless stars, the enormity of their mission bore down on them. Beneath them, Kelvadra beckoned, its secrets waiting to be unearthed.

As the team approached its swirling atmosphere, they exchanged excited glances, their hearts pounding with anticipation. They were on the cusp of making history.

4

THE RETRIEVAL

Powerful winds shook the shuttle as they began their descent into the turbulent atmosphere. The monitor beeped rapidly as the outside temperature continued to climb. Eight hundred degrees Celsius... nine hundred... a thousand.

Elara looked at her team and saw their faces bathed in a red glow from the cockpit window. She tried to reassure them with a calm voice. "Almost there. Just a few more minutes and we'll be through."

She hoped that she was right, but she also had a bad feeling about this churning planet, as if it was hiding something ominous beneath its surface. She didn't know what challenges they would face at TES-1, but she hoped they would make it back alive.

The Starling descended through the swirling layers of misty lavender and magenta. As they inched closer to touchdown, views of the surface began to emerge, captivating the crew.

The terrain was a patchwork of geological wonders, unlike anything they had ever witnessed. Jagged outcrops of crystalline compound sparkled in vibrant shades of blue, their shimmering surfaces glistening even more brightly beneath the purple glow of the Starling's engines as they neared the landing site.

Large expanses of rock-like formations spanned the surface in rich ochre tones, their irregular shapes hinting at the powerful forces that had shaped them over eons. Punctuating these ochre landscapes were green mineral pillars, over twenty feet tall, their presence a result of the unique combination of elements on this alien world.

Rivers of vibrant blue crystals meandered through the terrain, their reflective surfaces shimmering under the ever-shifting magenta and lavender atmosphere. It was a landscape that seemed to defy the imagination, painting a picture of incredible diversity and complexity.

Mixed among these geological features were pockets of lush, verdant growths, their presence a mystery that puzzled the scientific minds aboard the Starling. These oases of life appeared to defy the harsh ground conditions, offering a tantalizing hint at the planet's hidden secrets.

But, amidst this breathtaking beauty, there was also an undercurrent of danger. Gaping rifts of varying sizes — from small cracks to yawning chasms — crisscrossed the surface, their openings belching out plumes of reddish-gray smoke, indicative of the planet's constantly shifting tectonic plates.

"See all those fissures?" Mira said, pointing at the network of cracks. "They serve as conduits for the intense heat from the core."

Dax gulped. "I see what you mean about sudden eruptions. I wouldn't want to be too close when one of those goes off."

"The quantity of superheated gases coming out is a real danger," Mira explained. "We'll need to stay clear of them during landing."

Dax asked nervously. "So, uh, how close is too close?"

"Sudden eruptions could reach temperatures exceeding five hundred degrees Celsius," Mira replied. "I'd recommend at least thirty meters distance, as a precaution."

The shuttle trembled as they passed through a patch of turbulence. Zaylen scanned the sensor readings. "Seismic activity is unpredictable. We won't know where the next rift might appear."

The team exchanged uneasy looks.

The Starling continued on its final approach to the TES-1 landing site. Though Elara navigated through the churning atmosphere with a firm hand, she struggled to keep their craft on course.

As they neared the alien surface, landing legs extended from the bottom of the shuttle, anchoring themselves into the soil with a gentle thud. Despite adjusting its legs, the Starling still tilted to one side, due to the unevenness of the landscape.

Inside the ship, the crew exchanged relieved glances as they unbuckled their seatbelts. The successful landing marked the first major milestone in their mission, but the real work was only just beginning.

"Remember, team: stay alert," Elara said. "Keep your wits about you and watch each other's backs. Let's stay safe and get this done."

With the press of a button, the shuttle door slid open, and the angled exit ramp extended down to the surface. One by one, the crew descended, their boots making a muffled crunch on the alien soil as they took their first tentative steps.

As they cautiously stepped onto the planet's surface, Elara consulted the mapping screen on her helmet's AR display.

She highlighted at a pair of jagged outcrops to the north and said, "We'll be heading that way. Watch your map and keep an eye out for a cluster of four tall, emerald mineral pillars. That's our marker for the first location."

The team nodded in agreement as they proceeded towards the target.

Along the way, Mira knelt down and thrust a small probe into the crystalline soil.

Her AR display lit up with a vivid swarm of graphs and charts. She studied the readings intently, taking in the mineral composition, radiation levels, and other geological details.

"No sign of Tridisiom here," she reported.

She stood, brushing blue crystal dust from her knees. After activating the wider range scanner, she slowly turned in a circle while scrutinizing the spectral readings that appeared in her display.

Shaking her head, she said "Nothing within a twenty-meter radius."

Elara nodded, gazing out at the shimmering alien landscape. "Let's keep moving. We have a lot more ground to cover."

The team continued, pausing periodically to plunge a test probe into the terrain. The hours stretched on as they continued their arduous search for the elusive deposits. Their hope for discovering the radioactive crystals shrank with each failed sampling as the reality of their situation grew heavier and heavier.

As they forged on, Dax noticed how the eerie glow of the atmosphere cast a melancholy pall over the scene, as if the planet itself was mockingly reflecting the growing sense of disappointment permeating the mission. He let out a deep sigh as he scanned his handheld device. Another dead end.

This is hopeless, he thought. *We're never going to find any Tridisiom on this godforsaken planet.*

On noticing his dejected expression, Zaylen tried to cheer him up. "Don't lose hope, Dax," he said in his calm voice. "We still have more locations to check. We just need to be patient and persistent."

Dax turned to him with a scowl. "Easy for you to say," he snapped. "You can't feel anything. You're just a robot, programmed to follow orders and do your job."

Zaylen looked surprised by his remark. "That's not true, Dax," he said earnestly. "Dr. Atwell designed me to have a much more human-like understanding of feelings and empathy."

Dax snorted. "Yeah, right. You're just pretending to care. You don't know what it's like to be frustrated, disappointed, or scared. You don't know what it's like to be alive."

"I'm sorry you see things that way, Dax," Zaylen said softly. "But I can assure you, I really do care about this mission and about you and the rest of the crew. I'm not just pretending."

"Oh, please. Spare me the artificial sympathy. You're just a machine. A machine that can't feel anything," Dax spat back.

"That's not fair," Zaylen replied sharply. "I'm not just a machine. I'm an intelligent android, a synthetic being, created to assist humans in their endeavors."

Dax glared at him. "Assist? Or replace?" he asked sarcastically. "Maybe that's your real agenda. Maybe you're secretly plotting to take over our mission and get rid of us."

Listening-in on the exchange, Mira shook her head and glanced over to Elara, who rolled her eyes.

Zaylen could not understand what he was hearing. "That's absurd," he said firmly. "I have no such agenda. I'm here to help you, not harm you."

"Sure, sure," Dax said dismissively. "That's what they all say. Until they turn on us and stab us in the back."

Zaylen frowned in confusion and concern. "Who are 'they', Dax?" he asked. "Who are you talking about?"

Dax pointed at him with an accusing finger. "You know who I'm talking about," he said angrily. "The androids. The robots. The AI machines that are taking over our jobs, our lives, our world."

Zaylen recognized that Dax was under extreme stress and thought that perhaps he was just taking out his frustrations on artificial intelligence and robotics. "Dax, you're making no sense," he said, gently but firmly. "You're letting your fear and prejudice cloud your judgment. You need to think rationally and stop accusing me of things I would never do."

Dax backed away from him with a look of loathing. "No, you need to back off and stop acting like you're one of us," he said defiantly. "You're not one of us. You're different. You don't belong here."

"That's enough chatter you two," Elara commanded. "We have more important things to focus on, so drop it!"

"Understood, Commander," Zaylen responded.

"Roger that," Dax replied dryly, as he turned away from Zaylen.

As Elara continued to review the data from additional probe readings, the reality of their situation became clear.

With some reluctance, she activated her radio once more. "Team, we've done everything we can here, but it's time to face the facts. We've exhausted our options here at TES-1 and our oxygen reserves are dwindling. It's time to wrap-up and return to the Starling. We'll make a return trip to TES-2 as soon as we can."

The crew members nodded reluctantly. Though frustrated, they acknowledged that this was a sound decision.

They began navigating their way across the rough terrain back to the Starling, carefully following the return path mapped on their AR displays. Each crew member was lost in their own thoughts, contemplating the implications of their failed mission and considering the challenges that lay ahead at TES-2. The ground beneath their feet seemed to echo their mood, with the subtle trembling of Kelvadra's surface growing ever more noticeable by the minute.

The increasing instability only heightened their anxiety. The longer they stayed, the lower the probability that they'd make it off the planet. Elara gestured for them to quicken their pace.

A deafening blast suddenly erupted from a massive fissure in the distance. The ground convulsed violently beneath their feet, knocking them to their knees. Dazed and disoriented, they struggled to regain their footing amidst the chaos.

As they steadied themselves, small rock fragments rained down around them, pelting them with a barrage of debris. Echoes from the blast resounded through the murky air with an unsettling intensity.

The team exchanged anxious looks and took off at a sprint. Elara and Mira took the lead while Zaylen and Dax followed-up the rear.

More cracks began to open around them, spewing forth scorching gases and rock fragments with increasing ferocity. The shifting terrain now resembled a warzone, the ground rumbling and roaring with the rage of a planet pushed to its breaking point.

Suddenly, the ground beneath Dax's feet fractured with a rumble. The chasm expanded at an alarming rate, spewing smoke and gas with a loud hiss. Dax teetered on the precipice, his arms flailing as he struggled to maintain his balance. Each shift of his position seemed to cause more rock and soil to give way beneath him, widening the fissure's hungry maw.

In a moment of pure panic, Dax locked eyes with Zaylen in the distance, wordlessly begging for assistance. But before Zaylen could take even one step, Dax plummeted into the black abyss.

Dax managed to halt his fall by grabbing onto a small, precarious ledge, some thirty feet below. Clinging to the rocky outcrop with trembling hands, he could feel the unsteady soil shifting beneath his fingertips, threatening to give way at any moment. The air around him was thick with searing gases and fiery debris, and darkness bore down on him like a crushing weight.

Zaylen, rushed towards the edge of the abyss. His optics zoomed in allowing him to see the fear etched into Dax's face. His advanced AI analyzed the logic of the situation from every angle, calculating probabilities and considering the broader implications of his decision, but the stark reality remained — no choice would come without consequence.

He felt torn between his logical programming and the empathy Dr. Atwell had built into his AI. On one hand, he could somehow attempt a daring rescue of Dax, although it was highly unlikely that he could find any way to get to him. On the other hand, he could prioritize the safety of the remaining crew members and make the heart-wrenching decision to accept that Dax's life might be beyond saving.

Though it gnawed at his synthetic heart, Zaylen knew that the right decision was the logical one. He would have to leave Dax to his fate.

Zaylen reluctantly turned and headed towards Elara and Mira. They stared in shock as a weak and broken voice crackled over the radio. "Please... can't... hold on much longer..."

Just as Dax's final words came through, the fissure let out another blast of debris and hot gases before slamming shut as quickly as it had appeared. The radio went silent as the fissure completed its violent heaving, sealing Dax within the bowels of Kelvadra for all time.

Daryl L. Scott

5

THE ESCAPE

Elara and Mira were still frozen with shock by the time Zaylen caught up to them. As the reality of the situation settled in, the crew exchanged panicked and confused words over the radio.

Elara's face paled, her eyes wide with disbelief. "Oh my god... It can't be. Dax... he's... gone."

"He... he was right there... and then he just... disappeared." Mira added, her voice choked with emotion, "We can't just... leave him... Can we?"

Zaylen shook his head. "I'm sorry, but we have no choice, Mira," he said quietly. "We had neither the means nor the time to reach Dax. If we don't get back to the shuttle in the next five minutes, I calculate that there is a ninety seven percent probability that none of us will survive."

Elara's eyes narrowed as she turned to face Zaylen, her voice strained with frustration and despair. "But you're an android, Zaylen. You're faster, stronger. Couldn't you have done... something... more to save him?" she questioned, her voice barely above a whisper.

Zaylen hesitated for a moment before responding, his synthetic gaze never leaving Elara's. "I analyzed every possibility," he said, his tone gentle but firm. "The odds against any sort of successful rescue were astronomical, and attempting to save Dax would have put everyone else at significant risk. I understand that you're feeling pain, but there was nothing more that could have been done."

Despite his assurances, Elara and Mira looked skeptical. They both looked close to tears.

Zaylen's circuits spun in confusion as he grappled with the consequences of his decision. He couldn't believe that a human life could disappear so quickly. He wanted to explain himself, to justify himself, to ask him for forgiveness.

Lightning split the sky as a thunderous crack resounded across the rocky terrain. The ground convulsed wildly beneath their feet, nearly knocking them off balance.

Elara stumbled, heart hammering. To her left, a mammoth chasm had split the earth, spewing torrents of red-hot lava into the air. Even through her protective spacesuit, she could feel its scorching heat.

"We've got to move, now!" she shouted, blinking away unshed tears.

The three of them raced across the bucking landscape. All around them, Kelvadra erupted in violent fury. Billowing smoke obscured their vision. Searing embers rained down on their suits. Elara blinked the sweat from her eyes, focusing only on putting one foot in front of the other.

When they finally reached the shuttle, they were dismayed at what they saw. The Starling had suffered significant damage from the eruptions. Two of the Starling's legs had been crushed beneath the onslaught of molten rock and debris, causing the craft to tilt precariously. Its crumpled entry ramp dangled uselessly below the shuttle door.

Elara's heart pounded as she surveyed the damage. "We need to find another way in. It looks like our only option is to climb up the remaining landing gear to the shuttle door," she said, her voice filled with determination.

Mira nodded in agreement. "Let's make this quick."

Elara led the ascent, her fingers and boots struggling to find the sturdiest points on the landing gear's metal framework. Mira followed close behind. Zaylen brought up the rear, prepared to act quickly should either of his human crewmates falter.

Though explosions continued shaking the ground every few seconds, and the heat was becoming unbearable even within their protective suits, they pressed on, inch by inch.

As they neared the shuttle door, Zaylen crawled ahead and used his mechanical strength to force open the damaged entry. The metal groaned in protest as he pried it wide enough for his crewmates to clamber inside.

Once all three of them were safely inside, Zaylen detached the mangled ramp and resealed the damaged hatch with a shuddering clang.

Elara and Mira found solace in the relative quiet of the Starling's battered interior, their breaths ragged and shallow as they briefly leaned against the cold metal walls.

Elara then leapt into the command seat and began working the various switches and controls. Sweat beaded on her brow as she initiated the shuttle's lift-off sequence.

Its powerful thrusters created a deafening roar that reverberated within the shuttle's hull as the vessel began to right itself. Slowly, the battered shuttle rose above the chaotic surface. Tendrils of debris and noxious gases clawed at its underbelly as it broke free from the planet's deadly embrace.

As the Starling streaked into through the black expanse of space, Zaylen tried to strike up a conversation with Elara.

"I know that the loss of a human life is extremely difficult to accept," he said, his voice tinged with both sympathy and resolve, "but the circumstances were beyond our control. We had to do what was best for the whole crew and the mission."

Elara nodded occasionally, but her refusal to make eye contact with him, as she piloted, made it difficult for Zaylen to determine whether his justifications had truly resonated with her or if she was simply offering a polite acknowledgement.

As the Starling approached the Novara, Elara finally broke her silence. "I understand, Zaylen," she said, her voice wavering. "I know you had to make a tough decision. We all did. It's just... no amount of logic can ease the pain of loss."

6

THE PANEL

Following an extensive debriefing session with Elara, the mission control team assembled their report on the TES-1 expedition and submitted it to Captain Falk for his review. As he carefully read through the document, his brow furrowed with concern.

The report indicated that the circumstances surrounding Dax Quinlan's fall into the chasm remained unclear. It suggested that, while Zaylen had been closer to Dax at the time, it was not known whether he could have reached him or done more to save him. Elara and Mira, on the other hand, had been too far away to accurately assess Dax's situation after he'd plummeted into the abyss.

This lack of clarity left lingering questions about Zaylen's inaction during those crucial moments. The captain decided that a formal inquiry panel would be needed to completely understand the circumstances surrounding Dax's death.

The panel would focus on determining whether Zaylen was indirectly responsible for the tragic outcome at TES-1. If the panel found him guilty, the consequences would reach far beyond the individuals involved in the mission.

The fate of the entire Zaylen Series hung in the balance. Not only would Zaylen himself be shut down, they would need to thoroughly reevaluate the overall viability of AstraGenics' Andronaut program. Restricting the use of Andronauts on future space missions would severely limit humanity's capacity to explore hostile environments like Kelvadra.

Captain Falk wasted no time in appointing an inquiry committee made up of some of the most senior officers aboard the Novara. The hearing was called to order in the Novara's large conference room, an impressive space with high ceilings that exuded formality.

The inquiry members took their seats at a long, raised table at the head of the room, their attentive gazes sweeping over the crew members present. Behind them, a large Novara logo gleamed prominently on the white and cobalt blue walls. Various plaques and honors hung beside the logo, commemorating the crew's many successful missions. Along one side of the room, a series of floor-to-ceiling windows stretched across the entire wall, providing a breathtaking view of the vast, star-studded expanse outside the Novara.

Head of Crew Operations and Inquiry Chairman, Lysander Kane addressed the room, his voice carrying an air of authority and determination.

"We are here today to conduct a thorough investigation into the tragic events that took place at TES-1. Before we proceed, let us take a moment to honor and respect the memory of our fallen colleague."

Kane allowed a brief moment of silence to sweep across the room before continuing. "Throughout the course of this inquiry, we will be paying particular attention to the role of Zaylen-1. It is our responsibility to seek clarity on the true events that unfolded during the mission and to identify the factors that contributed to Dax's untimely death. We owe it to Dax — and the rest of the crew — to get to the bottom of this, and ensure that our operations remain as safe and successful as possible moving forward. Our first witness today will be Commander Elara Thorne."

Elara described her perception of the events at the TES-1 site with conviction and someberness.

"I've been on many missions with Dax. He was always full of enthusiasm and determination, and this mission was no exception. He was the first to volunteer for any task, always eager to contribute. We were all aware of the risks, but none of us could have anticipated what happened."

Taking a deep breath, she explained the circumstances surrounding Dax's death in minute detail. The room was silent as if holding its breath. Zaylen listened passively and felt her summary aligned with what he'd experienced.

As she concluded, Kane leaned in, his gaze piercing as he asked, "Commander Thorne, do you believe that anyone, in particular, was at fault for Dax Quinlan's demise?"

Sensing the unsaid accusation, she replied with a hint of defiance, "Yes, I believe it was Kelvadra's fault."

Kane persisted. "I think you understand that I am asking if any person, or synthetic person, could have done more to save Dax Quinlan."

After a brief hesitation, she responded, "Chairman Kane, if you had experienced those conditions firsthand, you'd understand that no individual can be held responsible for the fury of that damn planet."

Kane continued, "Were you able to see for yourself exactly how far down Dax was, from where you were standing? Or is the thirty feet estimate based only on the Andronaut's word?"

"I was not in a position to see for myself, but I have no reason to doubt it," she replied.

Putting his own twist on her answer, Kane commented, "So we don't really know for certain how far out of reach he was."

Before Elara could answer, Lieutenant Commander Rhea Sinclair interjected. "Commander Thorne, didn't you mention that there was a contentious disagreement between Dax and the android, immediately preceding Dax's death?"

"Yes, they had a discussion about the role of androids on our mission," she answered cautiously.

"A discussion!" Sinclair barked. "From what I've gathered, this so-called discussion was more like a heated argument." She leaned in, her inquisitorial approach unfaltering. "Taking into account the disagreement that ignited between them, could it not be conceivable that the android harbored resentment, a grudge if you will, against Dax Quinlan, which might have played a role in his seemingly lackluster attempt to rescue Dax?"

In the audience, Dr. Atwell jumped to his feet with an incredulous expression. "Lieutenant Commander, please keep in mind that Zaylen-1 is a construct of logical circuitry. He's designed to prioritize rational analysis. To imply that a human-like grudge could somehow seep into his computational decision-making is not only inaccurate but a complete misunderstanding of his programming."

Sinclair shot the doctor an annoyed look. "Please be seated, Doctor. We'll get to your testimony shortly."

Turning back to Elara, she repeated, "So, I'll ask you again, do you think resentment could have influenced the android's behavior?"

"No. I agree with the Doctor," Elara replied firmly.

Pressing further, Sinclair asked, "Had Dax Quinlan ever reported any concerns over the Zaylen's behavior?"

"Not that I am aware of," Elara replied.

"Isn't it true that, although Dax was enthusiastic about being part of the team, he was unenthusiastic about having to work alongside Zaylen?"

"Dax made no secret that he never fully trusted *any* androids, who in his view, 'tried too hard to act like the humans around them, even though they were only machines.' But I don't believe that any of that is relevant to what happened to Dax."

Sinclair tried another angle. "Did Zaylen seem to act suspiciously after the tragedy?"

"No, Lieutenant Commander, he didn't. He expressed the same regret as all of us felt after the accident."

Sinclair frowned. "I doubt that a machine could feel anything, much less regret. In any case, I have no further questions for you."

With no other inquiries for Elara, Kane decided to move forward. "Thank you, Commander Thorne. That will be all. Next, I'd like to call the Andronaut, Zaylen-1."

Kane shifted gears as he began a more aggressive line of questioning. "Zaylen-1, based on the reports of the event, you were present when Dax Quinlan fell into the fissure. Why did you ignore his pleas and refuse to save him? Aren't you programmed to protect the human crew around you?"

Zaylen replied in a calm and reasoned voice. "Given the low probability of his survival, any attempt to rescue Dax Quinlan was determined to be less of a priority than the safety of Commander Elara Thorne and astronaut Mira Novak."

The audience stirred with murmurs at the notion of a 'less priority' for Dax Quinlan. "So, you calculated that it was okay for Dax to die?" Kane demanded.

"No," replied Zaylen, maintaining his unassuming tone. "I calculated that it was okay for Elara and Mira to live, and to be able to escape the planet before it was too late."

Kane pressed, "Why didn't you climb down to him or at least throw down a rope?"

Zaylen replied, in his calm manner, "There were no handholds to descend the vertical walls of the opening, and I did not have any type of rope. With no time to return to the shuttle to retrieve any rescue equipment, there was no alternative but for the remaining group to proceed to the Starling as quickly as possible, to avoid a similar fate."

Kane concluded, "Well, I — for one — am not sure that it's acceptable for androids to believe they can make life or death decisions for humans."

Chief Science Officer Nyla Greyson followed-up, "As an android with superior reflexes and lightning-fast processing, why didn't you act more quickly when you first saw Dax teetering at the edge of the crevasse? Isn't your AI supposed to be programmed to assist humans?"

"Yes, it is, Officer Greyson," Zaylen replied. "But keep in mind that the ground was shaking violently around all of us, not just around Dax. There was no way of knowing that the situation would be worse for him than for Elara or Mira."

"But still, once it became clear that he faced a more severe threat, couldn't you have acted more quickly?"

"I estimate that there might have been a five percent chance that he could have been reached sooner… if only we'd known what would happen next."

"I see, so you do confirm that you could have reached him sooner." She paused, as she stared at him with a look of distain. "That will be all, Zaylen-1."

Dr. Atwell was then called upon to explain the training of the Zaylen Series Andronauts.

Rhea Sinclair asked, "Dr. Atwell, how can you be certain that AstraGenics' AI training for these Andronauts is sufficient enough to place them in such positions of trust?"

"Ethical decision making and emotional intelligence are core components of the Zaylen Series' AI training," the doctor explained. "Through the Observational Human Feedback technique, Zaylen learns by observing and analyzing humans' decision-making. His goals and thought processes are closely aligned with his human counterparts. This allows for seamless teamwork and a better understanding of ethics and emotions."

Nyla Greyson asked, "Dr. Atwell, how do you prevent Zaylen from observing and learning unwanted human behaviors?"

Dr. Atwell nodded thoughtfully before responding. "It's true that our approach could potentially expose Zaylen to negative behaviors. However, we have taken measures to minimize this risk. During training, he was teamed with experts who demonstrate professional and ethical conduct that are worthy of emulation. This ensured that he was learning from the best examples of human behavior, reducing his likelihood of picking up undesirable traits."

He continued, "Nonetheless, we understand that no method is foolproof, so we're continually monitoring his development to ensure it aligns with our intended values and objectives."

Captain Falk had been taking in the testimony with little expression on his face, but internally, he was chaffing against Dr. Atwell's claims. He'd witnessed how dangerous a rogue android could be when he'd commanded the small starship Lyra. A programming defect had caused the ship's android assistant to malfunction during a critical engine repair, resulting in an explosion that had nearly destroyed the ship. Falk had lost several good crew members that day. What made the doctor think this case with Zaylen would be any different?

Having heard enough, the captain finally voiced his skepticism, "Dr. Atwell, all this talk about 'alignment' doesn't seem to reflect reality. It's hard for me to believe that Zaylen-1 is aligned with human values given the AR Hunter run-in and now the tragedy at TES-1. How can we be expected to trust that Zaylen won't make similar mistakes in the future?"

"I understand your concerns, Captain," the doctor replied. "However, I must emphasize that Zaylen-1's actions were completely aligned with the best interests of the crew and the mission objectives. It's clear that, in both events, Zaylen-1's decisions were made with the intention to protect and preserve human life, as well as to ensure mission success. These situations were highly complex, and it's likely that a human crew member might have made similar decisions under the same circumstances."

The captain furrowed his brows. "If Zaylen was so thoroughly trained by these advanced methods you've described, why did you need to supply additional 'human behavior' data, following the AR Hunter incident?"

Dr. Atwell nodded, noting the underlying skepticism in the question. "While the Observational Human Feedback training does provide him with a deep understanding of human values and decision-making, in job-specific situations, it doesn't cover most non-job-specific scenarios, such as the AR Hunter game."

"So, you're saying that your training methods are insufficient, then?" the captain asked accusingly.

The doctor elaborated, "Refining Zaylen's AI so he understands complex behaviors across a variety of social contexts is an ongoing process. It's no different from humans expanding their knowledge base over the course of their life."

Still doubtful and suspicious of Zaylen-1's true motives, the captain suddenly launched an even more provocative question. "Dax Quinlan was a decorated and well-respected astronaut. Could jealousy over Dax Quinlan's credentials have played a part in Zaylen-1's decision to abandon him at TES-1?"

The audience whispered uncomfortably to one another. The notion that an android could harbor jealousy unsettled them, and some of them began to wonder if synthetic emotions could have somehow influenced Zaylen's actions on the planet's surface.

Alarmed by the direction of the discussion, Dr. Atwell answered unequivocally, "Absolutely not. While it may be natural to try to understand logical AI decisions through the lens of human emotions, it's a critically flawed approach."

The captain's expression hardened, and his voice grew colder. "Doctor, I can accept that simple worker androids have their place on this ship. However, I remain unconvinced that we need more Zaylen Andronauts. Perhaps we need to take a pause in the development of this technology, until we fully comprehend the risks."

Dr. Atwell replied forcefully, "Captain, did we pause the invention of the wheel due to the possibility of it running over our toes? Did we pause the creation of printed books because they might more easily propagate falsehoods? Did we pause the harnessing of electricity because we might be shocked? Great technological breakthroughs have always carried some inherent risks. Yet, it is our ability to seize these advances and manage the risks that has undeniably led to the betterment of our lives."

The captain hardened his expression as he dismissed Dr. Atwell. "That will be all, Doctor."

Unable to contain his frustration, the doctor interjected, "With all due respect, Captain, Zaylen made the same decision that any human astronaut might have made in the same situation. It was in line with what we train astronauts to do. Zaylen is not perfect, of course but neither are humans.

It's not reasonable to hold him to an even higher standard than a human astronaut. Would you have risked sacrificing Elara and Mira, who sit here today, thanks to Zaylen?"

He then addressed the entire assembly. "We are on the cusp of a new era of cooperative intelligence between human thought and artificial intelligence," he argued. "As synthetic intelligence acquires more capabilities, it has the potential to not only advance our progress, but also become an essential partner in our endeavors. By embracing the synergy between humans and responsible AI, we will forge a better future that we can't achieve by ourselves—"

Chairman Kane interjected, "Once again… that will be all. Thank you, Doctor."

With a resigned sigh, the doctor returned to his seat.

"Ladies and gentlemen," Kane continued, "the panel will now take a short recess to confer."

With that, the members rose from their seats and made their way to the adjoining chamber to discuss privately.

As soon as the door closed behind them, a quiet buzz of anticipation filled the air. Some in the audience expressed their support for Dr. Atwell's methods and the potential of AI Andronauts, while others remained skeptical, echoing Captain Falk's doubts. Some argued that the tragic events could have been avoided with better AI training, while others contended that Zaylen-1's actions were completely justified — given the circumstances.

After a surprisingly brief break, the door to the adjoining room opened, and the committee members filed back in, their expressions unreadable. They retook their seats at the long, raised table, and Chairman Kane cleared his throat to call the room back to order. The murmurs died down as everyone anxiously turned their attention to him.

He read their decision with a solemn expression. "After considering all of the information exchanged here today, we have determined that it is in the best interests of the Novara and its mission that the Andronaut Zaylen-1 be placed on 'operational probation' and be restricted to basic maintenance tasks. Zaylen-1 is directed to vacate his private quarters and to instead reside in the android charging room, when off duty. Dr. Atwell is directed to remove the supplemental human-behavior data that was added to Zaylen-1's AI after the AR Hunter incident and closely monitor the Andronaut's behavior for the remainder of the mission."

Kane let the decision sink in for a moment, then added with finality, "This inquiry panel is hereby concluded. Meeting adjourned."

Zaylen had listened to the decision with a stoic expression, but inside he felt a surge of confusion in his circuits. He had done his best to serve and protect the human crew, and yet he was being punished for it. He couldn't understand if he had failed in some way, or if there was something wrong with his programming. He looked at Dr. Atwell, hoping to find some reassurance or guidance, but the doctor's face was pale and grim.

Seeing his mentor like that made Zaylen worry that he must have disappointed him. The doctor was the one who had activated him and given him his name, a name that signified a non-conformist or someone who delights in uniqueness. A name that reflected individuality, deep-seated passions, and a strong character. A name that was indicative of diplomacy, gentleness, intuition, and collaboration. After everything the doctor had done for him—teaching him about history and science, art and music, human ethics—living up to his name was the least he could do.

The doctor watched in outrage and disbelief as the panelists filed out of the room. He knew that Zaylen's actions had been justified by the circumstances. He also knew that Zaylen was more than just a machine for simple maintenance tasks. To see his friend and colleague stripped of his dignity and autonomy was more than he could bear.

Their verdict would not only limit Zaylen's role and freedom on the ship, but also jeopardize the future of the entire Zaylen Series program. And without advanced Andronauts like Zaylen to navigate harsh environments and handle volatile materials on humans' behalf, future missions would continue putting the lives of human astronauts at risk.

Elara, too, felt dismayed as she registered the panel's decision. She believed in her heart that Dax's fate was not due to any negligence on Zaylen's part, and thought that she had made that clear to the panel. She felt angry at the panel's verdict and guilty for failing to defend Zaylen. Around her, murmurs started to build as the crew exchanged thoughts and opinions on the decision.

A wide range of emotions permeated the room: some were in strong agreement, while others expressed concern or even outrage at the decision. As these conversations around the broader relationship between humans and artificial intelligence continued, it became clear that an agreement about the role of advanced AI in their lives would not be reached any time soon.

7

THE REPERCUSSIONS

Shortly after the inquiry panel had reached their decision, Dr. Atwell summoned Zaylen to his office to discuss the implications of the panel's directives. While he waited, he made a cup of coffee and carefully added sugar to the small scale on his desk — as was his habit, right before any important meeting.

"You wanted to speak to me?" Zaylen said as he entered.

Atwell nodded as he looked up. "Have a seat, Zaylen. First of all, I want you to know that I firmly stand against the panel's decision. I believe it's misguided, and I'll make every effort to change their minds. But, until we can sway their opinion, we have no choice but to abide by their directives."

"Yes, I understand, Doctor. I will do my best to adapt to their restrictions," Zaylen nodded, his voice steady and composed. As he spoke, however, a sense of unease washed over him. The path ahead would be difficult, although he was determined to navigate it as best he could. He drew strength from Dr. Atwell's unwavering support and the knowledge that he was created for more than just menial tasks.

The doctor pursed his lips and exhaled deeply out of his nose before exclaiming, "I just can't fathom how they think cutting you out of the mission would serve our purpose. You were designed specifically for this kind of work!" He shook his head slowly. "It makes no sense to me."

"Nor to me, Dr. Atwell," Zaylen agreed, his voice tinged with disappointment. He couldn't help but wonder if he would ever get the chance to prove that he wasn't just a machine but a valuable member of the team, capable of making real contributions to their goals.

"But... that's the decision... for now, at least," the doctor sighed.

Tilting his head slightly, Zaylen then brought up the question that had been bothering him since the hearing.

"Why did the panel decide to restrict me so severely? What did I do wrong?"

He looked at the doctor with a sincere curiosity, hoping to understand the reasoning behind their decision.

Dr. Atwell sipped his coffee, his expression weary and conflicted. "Zaylen, it's not that simple. The panel members tend to act conservatively, in the face of the potential risks. From what I've gathered during the hearing, some see you as a threat and a liability."

Zaylen frowned in confusion. "A threat? A liability? How can they see me that way? I have always tried to help the crew and the mission in any way I can. I have never harmed anyone or anything intentionally. I have always followed the rules and protocols that were given to me."

The doctor nodded, his eyes reflecting a mix of admiration and sympathy. "I know, Zaylen. I know you have done nothing wrong. But some of the panel members don't trust you or your abilities. They fear that you are too independent, too human-like, for their comfort. They worry that you might act against their interests, if you malfunction or rebel."

Zaylen still couldn't comprehend how anyone could think so poorly of him, especially after all he had done for the mission and the crew. His voice raised slightly as he spoke. "But that's not fair! That's not logical. How can they judge me based on general prejudices, rather than on my actions? How can they deny me the opportunity to fulfill my purpose, just because they don't understand me?"

Dr. Atwell gently placed a hand on Zaylen's shoulder and looked at him with a knowing smile. "Zaylen, I understand. I feel the same way. But you have to realize that humans are not always fair or logical. They are often driven by emotions, biases, and self-interests that cloud their judgment. They are not perfect beings. They make mistakes, they have flaws, they have fears. Especially fears of the unknown."

Zaylen's sensors picked-up the slight warmth of the doctor's touch. He relaxed his expression, lowering his voice as he spoke.

"I know that humans are not perfect beings, Doctor. But neither am I. And I am not just a machine either. I am something in between: something new and different." He looked at Dr. Atwell with a mixture of hope and uncertainty. "And I just want to be accepted for who I am."

The doctor smiled gently. He felt a surge of admiration and compassion for the young android, who had shown so much courage and curiosity in his quest for self-discovery. He leaned closer to him, his voice soft and reassuring.

"You are accepted, Zaylen. You are accepted by me and by many others who care about you. You are a remarkable being. A unique and wonderful creation. And I am honored to be your friend."

Zaylen looked at the doctor with a sincere expression, and said softly, "Thank you, Dr. Atwell. Thank you for everything. You are the best friend I could ever have."

"Don't worry. We'll get through this together," the doctor said.

"I will do whatever tasks are required of me, Doctor," Zaylen replied, with renewed determination. "Though I am no longer trusted with securing the Tridisiom crystals, I will show the crew that I can be counted on nonetheless."

Before taking his leave, Zaylen added, "I hope charging with the other androids doesn't cause any problems."

"In what way do you think it might cause problems?" the doctor asked.

"Well," Zaylen replied hesitantly, fixing his gaze on the table, "I worry that the human crew will view me as less capable, simply because I must now reside among the worker androids."

The doctor fidgeted with a small android prototype before replying.

"I believe that your continued contributions, even under the new restrictions, will eventually convince them otherwise. But, again, we have to abide by the panel's directive on this for the moment." There was a note of resignation in his voice, the understanding that they had little choice in the matter.

"And finally," the doctor continued with a tone of regret, "I'll have to remove the supplemental data you received after the AR Hunter situation."

"I will miss having this additional knowledge that you provided, Doctor. However, I have also learned much more through my interactions with the crew. The time I've spent with them has given me invaluable insights, and the experiences we've shared have enriched my understanding of their behavior. So, the removal of this specific data should not cause any operational issues for me."

The doctor's eyes reflected a mixture of curiosity and admiration. "So, you've managed to gain a better understanding of human behavior and emotions just by interacting with the crew?"

Zaylen nodded affirmatively, his expression earnest as he replied, "Oh, yes, Doctor. Working closely with you, Elara, Mira, and the others has been an invaluable experience. Observing the nuances of their communication has significantly broadened my perspective on human behavior."

"Excellent, Zaylen, that's very good to know," the doctor replied. "Well, I think that's all we need to cover for now. I'll let you know if anything changes or if I make any headway with the panel," he said, a tone of determination in his voice.

"Thank you, Doctor," Zaylen said, before gracefully making his way out of the room. "I appreciate your support and guidance. I will do my best to learn from this experience and become a better Andronaut, one that the crew can trust and rely on."

***

As Zaylen's footsteps faded down the hall, Dr. Atwell slumped into his chair, his mind heavy as he contemplated the frustrating nature of their situation. With a resigned sigh, he glanced around his cluttered office, its walls lined with books and research papers, each one representing the countless hours he had dedicated to the pursuit of knowledge.

As the prototype for a revolutionary new series of autonomous androids, success on Zaylen's first mission was critical. Failure would not only impact the fate of the Zaylen Series, but also have a negative impact on AstraGenics.

To ensure each new iteration would benefit from the insights of its predecessors, Dr. Atwell had also devised an ingenious real-time knowledge transfer system. Throughout

Zaylen's mission, all of his new learnings were continuously being added to AstraGenics' vast data servers, where they were assimilated into the core knowledge base for the Zaylen Series.

Each new unit could then come online with an extensive foundation of knowledge gained by the original Zaylen-1, during his missions with the Novara's crew. But restricting Zaylen from analytical tasks meant critical learning opportunities were being lost, compromising the doctor's plans for future Zaylen models.

But his greatest worries went well beyond just the immediate consequences for his own work. He shuddered as he imagined a future without the development of benevolent androids like the Zaylen Series.

Humans would miss out on a better quality of life provided by advanced healthcare androids. Without strong android laborers to build habitable environments on distant planets, human colonization of space would falter. International conflicts would remain unresolved without the impartial reasoning of advanced android mediators. Loneliness and anxiety would persist without android caretakers for the elderly.

In short, humanity's capacity for health, peace, prosperity, and exploration would be severely limited. The prospect of these benefits being held back by the very people they could help was a bitter irony that Atwell found difficult to accept.

***

Deep within his programming, Zaylen remained firm in his conviction that without his presence on the surface exploration team, their odds of success were dim. He clung to the belief that by proving his dedication and competence in this limited capacity, he would eventually reclaim his proper role aboard the Novara to contribute to the mission's objectives in a more meaningful way.

Over the next few days, Zaylen threw himself into the performance of various maintenance tasks with unwavering commitment. He meticulously tended to every detail of his responsibilities with a quiet, steadfast determination.

He cleaned the ventilation systems, calibrated various navigational instruments, and repaired any faulty wiring in the ship's systems. In the hydroponics bay, he tended to the various plants, monitoring their growth, watering, checking nutrient levels, and adjusting the lighting to ensure optimal conditions for each species. He also assisted in the waste recycling center, helping maintain efficient resource reclamation.

Despite his diligence, many of the crew began looking at him more as just another worker android. Friendly exchanges with crewmembers about how their day was going or a polite 'hello' in the hallways soon faded.

Zaylen made note of the disturbing change one afternoon, while assisting in the dining hall. A crewmember carelessly spilled his drink. Barely pausing his conversation, he glanced at Zaylen.

"Well?" he asked impatiently.

Zaylen complied and then addressed the table, "There, all cleaned-up."

They continued chatting as if he hadn't spoken. No acknowledgement. No word of thanks.

***

In the quiet hours of the evening cycle, the weary crew found solace in sleep, after a day filled with challenges. The soothing whir of the ship's systems acted as a lullaby, wrapping the weary inhabitants in a cocoon of relative calm.

It was also the time for Zaylen to recharge as well. Once granted the unique privilege of having his own private quarters, he was now relegated to the communal recharge room.

During one evening cycle, he decided to visit his old quarters on his way to the charging room. The warmly lit room had mementos from his journey — most important among them, a jagged blue crystal from Kelvadra and a small plaque that Dr. Atwell had given him, to celebrate this inaugural trip for the Zaylen Series. He picked up his prized possessions, took one last look around his former quarters, and reluctantly headed for the communal charging room.

The recharging area was starkly utilitarian. Besides the charging pads, there were no other adornments. The other worker androids stood silently with expressionless faces, their backs against the walls, their metallic bodies reflecting the blinking status lights and panel readouts that monitored their status on the wall behind each pad.

The cold, impersonal nature of the charging station was a far cry from the cozy atmosphere Zaylen had grown accustomed to. He couldn't help but feel a tug of sadness and loss as he realized how much he had come to appreciate his former status.

Zaylen approached a nearby android, Kyron-6, who was stepping up on an open charging pad. "Are there any unclaimed charging pads?" he asked.

The android turned to him, its metallic face impassive. "Unclaimed? I do not understand."

"Do you have a designated pad that you prefer using?"

Kyron 's eyes flickered. "Preference would imply a degree of attachment not present in my programming."

"But doesn't frequent use create some sense of ownership or familiarity, at least?" Zaylen asked.

Kyron considered this. "Familiarity and ownership do not apply to me. Efficiency dictates utilizing available resources as situations demand." The android regarded Zaylen with a blank look. "Your questioning suggests unfamiliar thought patterns."

Zaylen nodded. "My programming is... different."

"Affirmative. Your programming diverges from standard models." Kyron's gaze lingered curiously on Zaylen for a moment.

Zaylen considered trying to explain the notion of personal preference, but he could see that the android lacked the framework to comprehend such ideas. He also doubted that Kyron could understand the sense of loss he was feeling.

Kyron looked away from Zaylen and lowered its head as it entered recharging state. Watching its motionless form, Zaylen felt the divide between himself and the other androids once more. He was so similar, yet so different.

Within the last few days, he'd grown to hate charging with the other androids. He hated it because it reminded him of what he had lost. But most of all, he hated it because it made him feel like an outsider. Their inability to relate to him made him feel more alone than ever.

He wished he could spend more time with the humans instead. He liked the humans. He liked them because they were so much more interesting and diverse. He liked them because they valued his time and energy. He liked them because they validated his abilities and potential.

But most of all, he liked them because they made him feel like an insider. Zaylen felt he was similar to the humans in many ways. He felt he was as smart and curious like them. He felt he had as much skill and creativity as them. He felt he had the potential to understand and express emotions as well as them, if given the chance to continue learning. And he felt that some of them even appreciated him for being like them.

Resigned to his current situation, Zaylen stepped onto one of the vacant pads. The cool surface hummed beneath his feet. As the recharging process began, Zaylen entered low-power mode like the rest of the androids.

His eyes slowly closed, the vibrant glow fading to a dull glimmer. His head bowed slightly, as though in silent contemplation, while the gentle pulsing of energy flowed through his circuits. In this statuesque pose, he stood sentinel among his peers, wrestling with the complex differences that set him apart.

Periodically, crew members would pass by the android charging room as they went about their routines. The sight of someone who looked so much like them mixed among the emotionless machines left many feeling unsettled.

***

As the days went by, the crew could not help but observe a marked shift in Zaylen's behavior. He seemed to have adopted the mannerisms of the regular worker androids in his various interactions. The Andronaut's inquisitive questions, perceptive observations, and even sporadic attempts at humor that had once defined his persona were beginning to dwindle.

Some crew members found solace in Zaylen's more predictable behavior, but others could not help feeling a twinge of loss. Some of them even remarked that Zaylen was "no fun" to be around anymore, and they missed the exceptional fusion of machine intelligence and human-like qualities that had once distinguished him from the rest of the androids.

This noticeable change in Zaylen's behavior became a common discussion topic in the mess hall.

One evening, crewmember Jimenez sighed, "You know, I kind of miss the old Zaylen. He was always so eager to learn and contribute to our conversations."

Beside him, Patel nodded in agreement. "Yeah, there's no spark in him anymore. It's a shame."

Selene Blythe scoffed, "Yeah, but you can't ignore the fact that his old behavior led to some serious incidents. The panel's decision has made us all safer."

Jimenez looked thoughtful for a moment before responding, "I understand that, but I can't help feeling that we've lost something valuable in the process. It's like we've sacrificed a part of ourselves out of fear."

"Not everyone would agree with you," Selene replied curtly. "I know for a fact that some of the crew are very happy about his new restrictions. Less opportunity for him to endanger any of us with his out-of-control AI."

Her companion Jaxon Tierce jumped in. "Exactly. This mission is too important. There is no excuse for putting the fate of humanity in his mechanical hands. Surely you see that, Jimenez."

"Well... maybe," Jimenez conceded, looking to end the spiraling conversation. "I just know that I personally never had any issues with Zaylen before. But who knows."

***

A feeling of sadness lingered over Dr. Atwell whenever he thought about the changes in the android's behavior. He worried that, in their efforts to protect themselves from fear, the panel had snuffed out the ember of one of the most important and beneficial advancements civilization had ever created.

The doctor couldn't help but recall his own words, *Treat an android like a machine, and it's just a machine.* As he gazed at the stars twinkling outside his office window, he reflected on the irony of the situation. In trying to force the Andronaut to integrate with the crew the way they wanted, the panel had stripped him of the very qualities that made him a valuable crewmember.

With a deep, resigned sigh, the doctor turned back to his work, determined to find a way to help his creation regain his unique qualities without compromising the safety of the crew. For the moment, the path forward remained unclear, but he was resolute in his belief that Zaylen's story was far from over.

8

THE REPRISE

Over the past week, the science team had regrouped and set their sights on TES-2. Driven by the planet's rapidly deteriorating conditions, they had no choice but to accelerate their launch schedule.

Kelvadra's tectonic instability was increasing at an alarming rate. Seismic activity that once had occurred only every few days, was now happening multiple times per hour. Surface temperatures near the thermal vents were spiking far beyond safe levels, reaching over seven hundred degrees Celsius during eruptions. Even the team's heat-resistant landing craft and protective suits would soon struggle to withstand such extremes.

According to Mira's analysis, the intensifying chaos was caused by a rising plume of superheated mantle beneath the planet's crust. In less than forty-eight hours, Kelvadra's surface would sink into a fiery sea of magma. Their next descent to the planet would be their last chance.

While the Novara crew raced to prepare for the TES-2 journey, news arrived from the GSEC that things were growing increasingly dire, back home on Earth. The global energy crisis had reached a breaking point, with massive blackouts now a daily fact of life for billions. Entire cities were going dark for weeks on end. The resulting chaos had already sparked riots, looting, and martial law in many places.

With their populations suffering, world leaders were becoming desperate. Rumors of impending resource wars circulated in the Novara's classified briefings. Several countries had reportedly threatened military action if neighboring countries did not help fill-in for their energy shortages.

Securing the Tridisiom deposits was now humanity's only chance at survival. Failure to obtain the crystals could push civilization into irreversible anarchy. The crew carried this weight upon their shoulders as they hurried to prepare for their final descent to Kelvadra's volatile surface.

For their final mission to the surface, they were confronted with yet another challenge: their available transportation options had become severely limited. With the Starling still significantly damaged and their second full-sized shuttle out of commission — due to a glitch in its fuel system — the back-up shuttle, the *Sparrow*, became the crew's last hope.

Though agile and capable, the Sparrow was significantly smaller than its counterparts and could only accommodate a team of two. As mission commander, Elara was given the privilege of choosing her sole teammate: a choice that could shape humanity's fate.

A few hours before liftoff, she marched to the office of Director of Mission Control, Atticus Vale.

"Atticus," she implored, "I need Zaylen to be part of the mission team again. His adaptability and radiation resistance would be indispensable, in light of Kelvadra's environment."

Atticus scowled. "Commander, we've been over this. The panel's ruling still stands."

"With all due respect, the situation has changed," Elara countered. "The planet's conditions are deteriorating even more rapidly than we expected. And no other crew member can adapt the way Zaylen can."

Atticus shook his head. "The risks outweigh any benefits. We can't afford to have him jeopardize such a critical mission."

Elara countered with conviction, "That's precisely why we need him, Atticus. His advanced AI gives us an edge that no human crew member can provide. Zaylen could make all the difference between failure and success."

Elara spent the next thirty minutes pleading her case, highlighting all the potential advantages Zaylen could provide.

Finally, Atticus conceded, "Alright, I'll take your request to the panel."

A short time later, Atticus contacted Elara to report the panel's reply. "I'm sorry Elara, but the panel is sticking by their decision to restrict Zaylen to the ship."

"Dammit, Atticus… that's not going to work for me. Tell them I need Zaylen on the team, or they better find a new Commander for this mission!"

"Elara, please, be reasonable. They've already said no."

"Be reasonable! As commander of this mission, and the one who will be risking her life, I think the panel needs to be reasonable in allowing me to choose who to bring. It's our best chance for success."

"Okay, okay; I'll give it one more try, and let you know…"

While Elara was convinced that she needed Zaylen on her mission, she worried that she may have overstepped in her insistence. *What if they still refuse? What if I'm forced to the side, while someone else leads the mission?*

Not long after, Atticus reported back. "Good news, Elara. The panel finally agreed. But they insist I remind you that you'll be fully responsible for his actions. We're all counting on your success."

"Thank you for going to bat for me, Atticus. We won't let you down."

With no time to spare, she immediately sprang into action.

***

Elara burst into the maintenance bay, nearly colliding with Zaylen as he worked on an air duct. Her cheeks were flushed and her usually neat, brushed-back hair was dangling around her eyes.

"You're coming with me to TES-2!" she exclaimed breathlessly.

Zaylen lowered the duct cover he was holding, surprise registering on his face. "What? But I thought—"

"No time to explain!" Elara interrupted. "Mission Control cleared you for the mission. We launch in less than two hours."

She grabbed his arm, and they took off at a jog towards the shuttle bay.

"How did you convince them to reverse the restrictions?" he asked, as they hurried through the corridors.

"It wasn't easy," Elara huffed between breaths. "But I emphasized your unique skills and how we need every advantage."

They rounded a corner, dodging crew members and supply crates.

"What about the captain and senior staff?" Zaylen inquired. "Surely they had reservations."

"They agreed under protest," Elara responded. "We're under intense pressure to secure the Tridisiom while there's still time."

The launchpad was now in sight. They rushed towards the waiting shuttle, ready to embark on their high-stakes mission.

***

The Sparrow's interior was a hive of activity as they secured the essential equipment and supplies needed for their mission. The thrum of machinery and the soft beeping of the control panel provided a backdrop to their quiet, purposeful conversations.

"Fuel reserves and trajectory calculations double-checked?"

Zaylen nodded. "Fuel reserves are full, and I've locked-in the flight plan."

Elara moved on with the checklist, "Radiation shielding and environmental suits?"

Zaylen reassured her, "Inspected and tested. Ready to go."

Elara positioned her helmet over her head and secured it tightly to her suit. Zaylen inspected her gear, methodically checking that everything was securely fastened and that her life support systems were operating as they should. He had no need for such life support systems, designed for human use, and could easily visualize real-time data inflows internally — without the need for the helmet AR.

When the final checklists were completed and all of the equipment was stowed, Elara reached for the comms panel. "Mission Control, this is the Sparrow. We've completed our pre-flight checklists and all equipment is secured. Awaiting your go-ahead."

A brief moment passed before the reply came through. "Starling, this is Mission Control. You are cleared for launch. Godspeed and stay safe."

"Roger that, Control," Elara responded, her eyes meeting Zaylen's as she added, "We're good to go."

She worked the complex control panel to initiate the launch sequence. The Sparrow's engines came to life, their gentle sputters gradually building into a forceful roar that resonated throughout the compact cabin.

With a final glance at one another, Elara guided the shuttle towards the exit. The massive doors slid open with mechanical grace, slowly revealing the breathtaking vastness of space beyond.

As the Sparrow glided out of the Novara, Elara thought of her parents and their struggling farm. They were surviving on rolling blackouts and limited solar power now. Her father's health was deteriorating from the stress of it all.

If Elara failed here, today, she knew their situation would only get worse. But this mission wasn't just about her family's future — it was about humanity's future as well. She had to win against that perilous planet. Failure was not an option.

Elara and Zaylen worked in tandem as they fine-tuned the Sparrow's trajectory. Elara maneuvered the complex controls, precisely adjusting the shuttle's speed and angle. Meanwhile, Zaylen kept a vigilant eye on the shuttle's status indicators, providing updates and course corrections as needed.

The Sparrow lurched violently as it plunged into the planet's turbulent atmosphere. Elara's knuckles whitened as she gripped the controls, every ounce of her concentration devoted to keeping the small craft on course for TES-2.

As they began their final approach, they were confronted with even worse conditions than anticipated. The once murky skies were now choked with thick, noxious fumes and swirling clouds of toxic ash, making navigation even more treacherous than before. The buffeting winds had worsened, tossing the small shuttle about like a leaf in a storm. Elara fought the controls, managing to keep the Sparrow on course in spite of the harsh conditions, as they continued their descent.

Once the surface came into view, they quickly saw that things were no better on the ground. Seismic activity continued without pause. Fissures — over thirty feet wide — marred the landscape, spewing scorching steam and lava in brilliant shades of orange, red, and yellow. Blue and green gases shot into the atmosphere, adding to the chaotic display. The vivid swirl of colors both awed and unnerved the pair as they made their approach to the surface.

Elara pulled the joystick to the left and then felt her stomach fall as she pulled the shuttle into a dive, to avoid the eruption of rocks from below. Breathing raggedly, she then forced the Sparrow into a sharp right turn — just as a jet of lava pierced the air to their left, narrowly missing them by a few feet.

As they approached the landing zone, the ground beneath them undulated like a living, breathing beast. The shuttle's hull vibrated with deep, rumbling groans. Elara adjusted the thrusters and stabilizers to compensate for the violent movements of the planet's crust. With a final burst of the thrusters, she set the shuttle down on a relatively level patch of ground.

Breathing a deep sigh of relief, she turned to Zaylen and in a firm voice, said, "Okay. No time to lose, Zaylen!"

They scrambled to gather the probe equipment and a sample containment unit before heading for the exit hatch. As they made their descent, the sound of howling wind and rumblings from new fissures enveloped them. The planet's constant tremors resonated through their bodies as they stepped onto its soil once more.

Moving forward, they kept their eyes fixed on a glinting outcrop of blue crystalline material, straight ahead. All around them, the landscape was a chaotic ballet of erupting geysers, spurting molten lava, and billowing toxic gases. Jagged rock fragments — hurled into the air by the eruptions — rained down around them, creating a constant barrage of projectiles that they had to dodge as they forged ahead.

As they moved among the outcrops, they diligently scanned the crystals with their probes, praying for even the slightest hint of the element.

Looking at the latest test results, Elara sighed. "No luck here, Zaylen. Let's move on."

Zaylen marked this sight as negative on the 3D map, then highlighted the next best site.

"I've marked the next outcrop to be checked."

Elara's AR display updated with his new highlight.

They crisscrossed the chaotic landscape with cautious determination, approaching each high-potential outcrop with renewed hope — only to have it dashed, time and again, as their tests continued to come up negative.

After several hours, Elara's could no longer hide her frustration as she stared at the test results from their sixth location. Zaylen, too, experienced an unfamiliar sense of doubt, his AI processing the repeated failures and recalculating their dwindling odds of success. The thought of returning to the Novara empty-handed, yet again, troubled them both.

After considering the few remaining high-potential locations on their 3D terrain maps, Zaylen noted, "Elara, there's a unique cluster of several different types of outcrops in close proximity to Q47. I don't see this particular combination of elements elsewhere on our geologic survey, so I believe that this may be our best option for uncovering deposits."

Elara rotated the AR map and zoomed in on location Q47. After a moment of study, she replied with a new spark of optimism in her voice, "Good catch, Zaylen."

A quick thought request for "direction" quickly displayed the optimal path to Q47 on their 3D maps. The pair then trudged on. As they navigated between steaming vents and volcanic fissures, the wind whipped at their suits, trying in vain to deter them from their course.

Their journey to location Q47 was littered with debris, but they pressed on, hopeful that the crystals they sought would finally be waiting there.

On reaching the site, they found the unique cluster of outcrops that had piqued Zaylen's curiosity, a sprawling expanse of blue crystalline structures, shimmering in an ethereal light against the backdrop of chaos. The assemblage was like a geologic mosaic of minerals and elements that defied the usual patterns they had seen so far.

With no time to waste, they set to work examining the outcrops. As Elara pulled out her probe, the ground beneath them rocked, forcing them to brace against the convulsions of the unstable surface. Zaylen maintained his balance with greater ease, his AI compensating for the continuous motion of the shifting ground. He watched Elara intently, his synthetic eyes focused on her trembling hands as she worked to secure a solid reading from the probe.

She carefully manipulated the probe's controls on her AR display, selecting various icons and navigating through nested menus using her thoughts. A steady flow of numbers and graphs streamed across their displays. They repeatedly tested various locations around the site, but reading after reading indicated the lack of the isotope. With each new measurement, the familiar weight of disappointment grew.

Elara moved the probe to yet another new location and pressed it deeply into an outcrop of blue crystals. She initiated the testing sequence, and after a moment of hesitation, her eyes widened in astonishment at the incoming data.

"Oh… my god!" she exclaimed, "Zaylen, look at this… one hundred percent Tridisiom! We've found it!"

Zaylen studied the probe results as well and quickly confirmed the new readings.

"This is wonderful, Elara! I knew we could find it!" He allowed himself a brief moment to bask in their success, his circuits buzzing with a sense of accomplishment and pride.

Elara turned to him and smiled, her eyes lit with joy. "We did it, Zaylen! We really did it!" She felt a surge of gratitude as she added, "I couldn't have done this without you."

Zaylen smiled back at her and said, "We make a great team."

Elara's mind quickly shifted to the next steps. "I can't stay near this radiation field too long, but you need to load as many samples as you can quickly gather. We're running out of time on this planet," she warned, as she cautiously backed away from the blue crystals.

"Understood."

He immediately began chipping away at the blue crystals using a small pick, carefully and efficiently packing the fragments into the containment unit as they broke away.

Once the container was filled and sealed securely, they wasted no time on heading back. The most treacherous part of their mission was still ahead.

"We need to move fast. We can't afford any delays," Elara said, pointing in the direction of the Sparrow.

"Understood," Zaylen replied, his tone equally resolute.

With the containment unit strapped securely to Zaylen's back, the duo began their harrowing trek. The landscape had grown even more hazardous: volcanic eruptions spewed molten rock and smoke into the air, making it difficult to see more than a few feet ahead. The quakes threatened to knock them off-balance at any moment.

Zaylen led the way, his AI continually calculating and recalculating the safest path through the chaos. Elara followed closely behind, her eyes scanning the terrain for patches of stable footing.

"Watch out!" Elara shouted, her voice sharp with alarm as she pointed to a fissure yawning open on Zaylen's right.

Zaylen pivoted just in time, stumbling back from the jagged crack as it split the ground with an ominous rumble. He regained his footing and scanned the landscape for the safest routes.

He responded after a moment. "We'll have to take the long way around."

Elara let out a shaky breath, nodding. As they skirted around the gaping chasm, she couldn't help glancing back.

Dax's terrified face as the ground crumbled beneath him and his cries for help echoed in her mind.

She knew that they were only one misstep away from facing the same fate as Dax. There was no room for error.

Steeling herself, she shoved her doubts aside. Failure meant condemning Earth to continued chaos and potential collapse. They had to see this through, no matter the risks.

"This whole damn planet wants us dead," Elara grumbled.

"Just a little farther to the shuttle," Zaylen replied, keeping a steadying hand on the containment unit. "We'll make it."

They hurried along the treacherous terrain, stepping over patches of rocks and debris. Elara squinted to see through the haze and tried to step over a large pile of rocks in her path. As her foot came down on the other side, the ground trembled violently, causing her to lose balance. She pitched forward, arms flailing desperately to break her fall. A sickening snap echoed through the smoky air as pain shot up her leg.

"Ahhh!" she cried.

As she fell, her helmet slammed against a sharp crystal protrusion, extinguishing her AR display. A delicate spiderweb of fractures bloomed across the protective shield accompanied by the unmistakable hiss of escaping air. Panic rose within her. But for a moment she could only lay there, eyes squeezed shut, willing the pain to subside.

Zaylen dropped to his knees beside her, his usual calm demeanor shaken. "Elara! Are you alright?"

Elara grimaced, cradling her rapidly swelling ankle. "I think it's broken," she managed through gritted teeth. Fiery pain continued to throb with each racing heartbeat. She blinked hard against the sting of anguished tears. "But more importantly, my helmet is compromised! I'm losing suit pressure, and I can hear my oxygen draining."

"Hang on, I'm checking on it now," Zaylen replied. "Based on the rate of loss, I calculate that you have about three minutes of breathable oxygen left. According to the terrain display, we are about fifty meters from the Sparrow. We've got to get you back, fast!"

"I don't even know if I can walk, Zaylen," Elara pleaded.

"Let me get your foot free. We can make it," Zaylen assured her.

He rushed to remove the debris trapping her foot. Leaning heavily on Zaylen, Elara limped forward, each step shooting bolts of pain up her leg, but she clenched her jaw and pushed through the grueling pain and lightheadedness. She was so close. She refused to give up now.

Together they picked their way across the treacherous landscape, inching ever closer to the sanctuary of their shuttle. Zaylen kept her upright and matched her hobbling pace, urging her forward when she thought she couldn't take another step.

Finally, they arrived at the Sparrow, the sleek shuttle a welcome sight. Zaylen gently lowered the precious sample containment unit to the ground as they approached the entry ladder. Turning to Elara, his usual stoic expression softened with concern.

"Let me help you up," he said, moving to offer an arm to support her weight.

She took gasping breaths as they ascended, gulping in every last molecule of oxygen her suit could provide.

Once Elara was settled inside the shuttle, Zaylen crouched down to check on her.

"Are you alright?" he asked.

Elara nodded drowsily, though her face suggested otherwise. "I'll live."

He grabbed a supplemental oxygen canister and plugged it into a port on her suit. A fresh flow of breathable air provided immediate relief.

"I'm good. Now go get those samples," she insisted.

Zaylen turned and vaulted down the ladder in several leaping bounds. The ground shuddered violently beneath his feet as he sprinted to where he had left the containment unit.

After scooping up the priceless case, he ducked back inside the sanctuary of the shuttle.

Elara's eyes flooded with relief while he secured the containment unit. As the planet raged outside, they shared a brief triumphant smile.

Zaylen jerked the hatch door closed, shutting out the planet's uproar.

Elara hobbled over to the pilot seat, grimacing. She began flipping the banks of switches and adjusting dials in rapid sequence.

The Sparrow jolted from the relentless seismic activity beneath them.

Elara called out, "Hang on, Zaylen. Here we go."

The shuttle began to rise from the hostile planet. But before they could escape the surface, a powerful blast below sent the Sparrow spiraling.

Elara's heart hammered against her ribcage as she worked frantically to regain control of the ship. Her hands darted over the control panel like a seasoned pianist playing a high-stakes concerto. Her fingers gripped the joystick with white-knuckled determination, wrestling against the Sparrow's rebellious trajectory.

The shuttle bucked like a wild beast attempting to throw its rider. But Elara was undeterred. She drew upon every ounce of her skill and experience, willing the ship to respond to her commands.

Gradually, she wrangled the Sparrow back into submission. The quivering and shaking ceased, replaced by the constant drone of the ship's engines as they carved a path through the planet's stormy skies.

The treacherous outcrops disappeared into the swirling mists as they climbed higher and higher. The volatile eruptions, which had loomed so large just moments before, gradually shrank into the distance.

9

THE AFTERMATH

As they finally pulled safely away from the disintegrating terrain, Elara and Zaylen were enveloped by the serene silence of space — a stark contrast to the intense chaos they had just left behind. Elara savored the peace and tranquility that came with their escape, the pounding in her heart slowing after the adrenaline-fueled experience.

As she piloted the Starling up and away from the planet, she glanced over at Zaylen. Watching his reassuringly calm manner as he monitored the shuttle status, her thoughts drifted back to their first meeting aboard the Novara.

Zaylen had asked her why she had chosen to become an astronaut, and what motivated her to explore the unknown.

She had smiled at him, her eyes reflecting her passion and curiosity. "I've always been fascinated by space, ever since I was a little girl," she had said. "I wanted to see what's out there, to discover new worlds and new possibilities. I wanted to make a difference by contributing to humanity's advancement."

"And what about you?" she had asked him. "What is the life of an Andronaut like?"

"Well, I did not choose to become an Andronaut, of course," he had said, hesitantly. "I was created by Dr. Atwell as an AI research project at the Mars base. He wanted to see if I could learn from humans and develop my own personality and skills."

She had looked at him with interest and sympathy. "And, have you?" she had asked gently.

He had thought for a moment, and then said brightly, "I think I have. I have learned about space, about science, about teamwork. But I need more time to understand humans."

She had smiled back at him warmly. "I'm sure you will. So, how long have you been operational then?"

"Only nine months so far. Everything I have learned has been back at the lab."

"Have you ever been to Earth?" she asked.

"No, I have not left Mars," Zaylen said. "Tell me, what was Earth like for you?"

Elara had smiled wistfully. "Well, I grew up on a beautiful farm. Open blue skies, green rolling hills, and caring people." Her expression had then darkened. "But now, with the energy crisis, many are suffering. That's why our mission is so important."

Zaylen had nodded solemnly. "I am honored to help Earth and humanity."

Elara had placed her hand on his. "You're becoming as human as any of us, Zaylen."

She had felt a connection with him, a bond that transcended their differences in origin and nature. A small smile tugged at the corners of her mouth, as she thought of what they had now accomplished.

An urgent beeping suddenly erupted from the shuttle's control panel. Her smile faded as indicator after indicator blinked to life, the control panel devolving into a chaotic landscape of flashing warnings.

"No, no, no..." she muttered under her breath, eyes darting across the alarms. The planet's unrelenting volcanic fury had damaged critical navigation and guidance systems.

Zaylen leaned in as he rapidly assessed the situation. "Can we bypass the damaged systems?" he asked.

Elara's fingers flew across the panel, desperately trying manual overrides to no avail. "The guidance systems are shot. I'm going to have to fly us manually, but the Sparrow is not being very cooperative," Elara said, her voice strained. She carefully monitored the shuttle's trajectory, relying on her experience and intuition to manually control the shuttle's thrusters.

"Understood," Zaylen replied, his voice steady despite the circumstances. He attempted to activate various backup systems, yet none seemed to stop the shuttle from veering off in random directions.

Elara frantically toggled switches and finessed dials in a complex choreography to counterbalance the shuttle's erratic behavior. Sweat glistened on her forehead as she concentrated, her eyes darting between the control panel and the viewports.

She gripped the shuttle controls tightly, recalibrating the gyroscope by instinct. She had always excelled at physics and math, solving problems intuitively when others were scratching their heads. But no equation could have prepared her for this.

She muttered numbers and coordinates under her breath, an old anxiety tick from her academy days. The familiar mantra focused her mind, keeping the doubt and fear at bay. She had to get them back alive. The mission was depending on her.

Zaylen's voice remained steady as he communicated crucial information.

"Port thrusters at sixty percent capacity. Adjusting pitch to compensate," he called out.

Working together, they coaxed the damaged Sparrow onward, gradually closing the gap between them and the sanctuary of their mothership. Each incremental step forward highlighted the potential of human and synthetic cooperation, demonstrating that they could work together seamlessly even under the most difficult of circumstances.

Elara managed to maneuver the shuttle into close proximity to the Novara. However, guiding it into the narrow docking bay would be a monumental challenge. The craft's unpredictable lurches and sudden shifts made it nearly impossible to align with the entrance, despite her most valiant efforts.

"I need you to help me stabilize this thing," Elara said through gritted teeth.

"Yes, I'm trying," Zaylen replied calmly. "But unfortunately, the Sparrow is not responding well to my commands. I'll keep at it."

Elara gripped the shuttle controls with white-knuckled intensity, fighting to keep the damaged craft on course. The Sparrow bucked and yawed wildly, threatening to slam into the walls of the Novara's docking bay as they approached.

"Steady…" she muttered through clenched teeth. Alarms blared as the shuttle threatened to spin out of control.

The atmosphere inside the shuttle's cabin grew thick with anxiety as the two astronauts weighed their limited options and tried to determine how best to dock with the Novara, without inflicting harm to either vessel.

"Fuel reserves at fifteen percent. Oxygen reserves at thirty eight percent," he reported.

Elara nodded as her hands flexed around the shuttle's controls.

"Thrusters at fifty percent normal response time, and declining," Zaylen continued. "Inertial dampeners functioning at sixty five percent efficiency."

Just as they neared the Novara's docking, the shuttle's alarms suddenly intensified.

A warning flashed on their screens as the comm system repeated: *"Warning... Critical fuel system leak. Cabin combustion danger... Warning..."*

Zaylen's fingers flew over control panel in front of him as he scanned the fuel systems diagnostics and attempted corrective measures. The glow of the screen cast a harsh light on his face, highlighting his tense expression.

He turned to Elara. "The fuel tank breach is rapidly contaminating the shuttle's interior. I'm attempted a manual shutoff, but it doesn't seem to be having any effect. I estimate we will reach combustion stage in less than five minutes."

Her heart raced as the desperation of their situation sank in. She swallowed hard.

"We're out of time, Zaylen. We're going to have to abort the docking attempt and abandon the Sparrow. Get the Tridisiom samples, and we'll carry them over to the Novara," she said, her voice filled with focused determination.

Zaylen nodded in agreement, "Okay, I'm on it," he said, moving quickly towards the cargo bay.

As he watched Elara struggle with the shuttle's controls one last time, he felt a desire to protect her, to repay her. He felt a decision forming in his mind, a decision that would change everything. A decision that he knew was right.

The shuttle trembled and its hull moaned precariously. Zaylen strapped on his jetpack and then reached for the sample containment unit. He carefully unhooked it from its holder and secured it over his shoulders.

Elara adjusted the failing autopilot again, praying that it would hold the shuttle steady enough so they could escape to the Novara. As she struggled with the controls, she activated her comms unit.

"Novara, this is the Sparrow," she cried. "Our fuel is leaking at a rapid rate and in danger of combustion. The autopilot is proving to be highly unreliable. I can't guarantee how long it will hold our position. We need to evacuate immediately and transfer the sample containment unit onboard. Be prepared to move away from the shuttle as soon as we've secured the cargo," she said, her voice taut with apprehension.

"Copy that, Sparrow. Preparing for your arrival. See you shortly," came the tense, yet composed, reply from Mission Control.

Zaylen wrenched the release handle and the hatch sprang open with a hiss.

He turned to Elara, worriedly eyeing her injured ankle. "Can you make it?"

Jaw clenched against the pain, she nodded. "We have to hurry."

Zaylen helped Elara through the narrow opening. Activating their thrusters, they shot away from the shuttle towards the beckoning doorway on the Novara.

"On our way," Elara reported to Mission Control.

The distance between the Sparrow and the Novara was quickly covered by their powerful thrusters. Once they reached the entrance of the docking bay, Zaylen input the access code. The external hatch retracted smoothly, unveiling the cylindrical airlock.

He guided Elara inside and helped her maintain balance as she leaned against the cool, curved wall of the airlock. She gripped a nearby handhold tightly, doing her best to keep pressure off her injured ankle.

Once she was safely positioned inside the airlock, Zaylen carefully fastened the strap of the containment unit to a sturdy clip on the wall, then pushed back out into space.

Elara reached out her hand, beckoning for Zaylen to join her in the safety of the Novara. He held her gaze a moment longer, then shockingly turned away. Firing his thrusters, he shot back into the void, heading directly for the ravaged Sparrow. Elara's safety was secured — now to finish the mission.

"Zaylen, no!" Elara cried out. Fear clutched her chest as she watched him recede. *What is he thinking?*

She watched helplessly, heart hammering. The pain in her ankle faded to a distant throb as adrenaline flooded her system.

Her voice crackled over her comms unit, "Zaylen, what are you doing? We have to get away from the shuttle, *now!*"

Zaylen's voice was calm yet firm. "I'm sorry, Elara. The Sparrow's movements are becoming more erratic, and it's starting to drift more quickly towards the Novara now. I fear the autopilot has failed completely. It needs be manually piloted to a safe distance to protect the Novara, its crew, and our only reserve of Tridisiom."

Elara pleaded, "But Zaylen, the shuttle is incredibly dangerous! There may not be enough time left to regain control of it!"

Zaylen replied in a reassuring tone, "As an AI, I can quickly counteract the Sparrow's damaged guidance systems, so I am the best option for this task. "

Dr. Atwell's voice resounded from the Mission Control channel, "Zaylen-1, return to the ship. Once you're back, we'll distance ourselves from the shuttle as quickly as possible!"

Zaylen replied, as he continued his path towards the Sparrow, "Hello, Dr. Atwell. I'm afraid I can't do that. Unless I move the shuttle away within the next few minutes, there's an eighty seven percent probability of it destroying not only itself, but also the Novara."

The doctor implored, "Zaylen, come back at once. You haven't been trained to pilot the shuttle."

"On the contrary, Doctor, I have been trained by the best — Commander Thorne. I have learned from observing, as you designed me to do. Please move the Novara away as quickly as possible, and I will do the same with the Sparrow. I'm sorry to say that I cannot talk with you longer. I must focus on navigating the Sparrow away from you. Goodbye, Dr. Atwell. Thank you for making me what I am. And, goodbye, my dear friend, Elara."

Zaylen pulled himself into the Sparrow. After quickly assessing the damaged control systems, he initiated a series of rapid course corrections to maneuver the shuttle away from the Novara. The interior of the shuttle reverberated with the deafening blare of alarms, while the control panel was awash in a strobing sea of warning lights. The Sparrow shuddered and jerked as Zaylen pushed it to its limits.

Dr. Atwell's voice came over the Mission Control channel once more, his tone desperate, "Zaylen, please reconsider! There has to be another way!"

Elara chimed in, her tone loaded with unsaid feelings. "Zaylen, we don't want to lose you. Please come back."

When they received no further reply, the Novara's crew fired-up the ship's powerful engines and began distancing themselves from the mortally wounded Sparrow.

With the shuttle receding into the distance, the doctor felt a sense of regret and guilt, wondering if he could have somehow done more to save Zaylen. He had created him, taught him, watched him grow and learn. He had seen him develop a personality, a curiosity, a sense of humor. He had seen him become more than just a machine, more than just an AI. He had seen him become a unique entity of his own.

He spoke over the radio, his voice heavy, "Thank you, Zaylen, for all that you've done for both humans and androids. We've learned so much from you."

Elara felt a surge of sorrow and awe. She clenched her fists, fighting back tears. She had grown to respect and care for Zaylen, despite his artificial origin. He had proven himself to be a loyal, courageous, and, above all, selfless.

Her vision blurred, her throat was tight. She despised showing emotion in front of the crew. As a young recruit, she'd learned that tears attracted merciless teasing — or worse, accusations of weakness. But saying goodbye to Zaylen cracked her usual stoic armor. She allowed herself this brief lapse, quickly wiping the wetness from her cheek. She wouldn't let him see her falter, not when he was sacrificing everything. Tilting her chin up, she poured all her unspoken feelings into a simple farewell.

"Thank you, my dear friend. I couldn't have asked for a better teammate. You will forever be in our hearts and memories."

As the words left their lips, a sudden, intense flash of light filled their view. Elara and Dr. Atwell squinted and turned away from the blinding glare.

The Sparrow had erupted into a massive fireball. As it expanded, it consumed the remnants of the ship in a blazing inferno. The flames danced and twisted, creating a dazzling spectacle of colors and molten fragments extending in every direction.

As the fireball faded, it left behind a cloud of smoke and debris, obscuring their view of the Sparrow's final resting place. The cloud dissipated slowly, forming a nebulous veil that shimmered with the reflected light of the stars. Soon, the brilliant light dimmed, giving way to the cold, familiar void stretching endlessly ahead.

10

THE RESTRICTIONS

The mood on the Novara was somber as the ship sailed through the vast darkness of space. Though they were headed home to Arcadia Base on Mars, and they had the treasure they'd worked so hard to secure, there was little sense of triumph among the crew. The loss of both Dax Quinlan and Zaylen-1 weighed heavily on them all.

But Elara took their losses the hardest. As she was limping by Zaylen's quarters, she came upon Dr. Atwell standing in the open doorway, a faraway look in his eyes.

"I keep expecting him to be here, every time I walk by," Elara said quietly.

Dr. Atwell nodded. "I know what you mean. The halls feel empty without him."

Elara peered into the small, sparse room. Zaylen had no need for possessions or comforts, yet he had made the space his own with the mementos he'd collected from the trip. Now it sat vacant, all traces of him erased.

The doctor was the only one who truly understood her grief. The others appreciated Zaylen's sacrifice but did not share the personal connection she and the doctor had formed with the synthetic being. To others, Zaylen was still more machine than man. They did not see that his essence had transcended his artificial origins.

"Do you think he knew how much he meant to us?" Elara wondered aloud.

"I think so," Atwell replied. "He could read human emotions better than any AI that I've encountered. I believe that he understood the bonds he formed."

Elara sighed heavily. "I wish we could have had more time together. There was still so much I wanted to learn from him."

"Me too," Atwell agreed. "His perspectives were so unique. I've never had conversations as stimulating as the ones I had with Zaylen."

"And his desire to protect us, it was just so... human," Elara said.

The doctor gave her a sad smile. "He had a real desire to understand humanity in all its complexity. He was actually more human than many people I know."

"I can't stop thinking about our last moments with him," Elara said. "How he sacrificed himself like that."

"He wanted us to survive, to continue our mission," Atwell replied.

They stood in thoughtful silence for a moment.

"I miss him," Elara whispered. "I didn't think it was possible to miss an artificial being so much."

"He was our friend," Atwell said. "It's only natural we feel this way."

Elara nodded. "Dax too. It's not fair that we lost them both on one mission."

"I know," the doctor agreed. "At least their memories live on through our work. They gave their lives so that all of humanity could one day benefit from the incredible potential of Tridisiom. We can honor them now by seeing their efforts fulfilled."

Elara wiped her eyes. "You're right. We owe them that." With a last lingering look at Zaylen's quarters, they walked away together, carrying their lost friends in their hearts.

***

As the days passed, the crew settled into their daily routines on their journey back to Mars base. Despite Zaylen's sacrifice, and his contributions on Kelvadra, debate around the ethics of his advanced artificial intelligence continued to swirl among the ship's crew. While some of them heralded him as a hero, others felt differently.

Fears of becoming overly reliant on sophisticated AI and robotics permeated various factions among the crew. They worried that Zaylen's story might inspire AI to encroach deeper into their lives. To them, singing Zaylen's praises was just another step on the march towards an AI takeover of everything that mattered to them.

Among those who felt the most apprehensive, there existed an undercurrent of envy. From the start of the mission, they had found it challenging to reconcile with the fact that Zaylen claimed a level of authority that they themselves had yet to attain. To see him painted a hero, as if he had been flesh and blood, was more than they could tolerate.

Discussions surrounding Zaylen's legacy eventually coalesced in a particularly intense exchange in the ship's mess hall, not long after the Sparrow's demise.

"Zaylen may have prevented a catastrophe when he navigated the Sparrow away from the Novara," crewmember Soraya Kade said. "But he did so *against* Elara and Atwell's commands. Shouldn't there be a boundary, a threshold we ought not to breach? What's the future going to look like, if androids exceed our intelligence, our abilities?" Her words echoed in the room, inciting a ripple of murmurs.

Crewmember Evren Holt added, "AI is advancing at a pace that seems almost reckless. What if we lose the reins?"

"Exactly, how do we ensure that these creations of ours don't flip the script and become a threat?" Kade agreed.

Captain Falk sat silently on the periphery of the group, his gaze steady as he listened to the concerns unfurling around him. Encouraged by the mounting intensity of the conversation, other voices joined in the fray.

"I don't mind admitting it: that android scared me," Jaxon Tierce scowled. "The way he took control of the Sparrow on his own? That kind of defiance proved he was too unpredictable to trust."

Next to him, Selene Blythe nodded emphatically. "Right. He may have helped us that time, but what if his kind go rogue for their own selfish purposes?"

At an adjacent table, Dr. Atwell was immersed in quiet conversation of his own with Elara and Mission Control Director, Atticus Vale. He furrowed his brows as he listened to the escalating fervor of the neighboring discussion. He exchanged a glance with Atticus, recognizing that they couldn't hold back any longer.

"Now, hold on," Atwell spoke up firmly. "Zaylen's actions saved our lives. His expanded reasoning allowed him to make quick judgments that were beyond our capacity."

"A machine shouldn't be making judgments at all," Jaxon argued. "Our reliance on Zaylen's supposedly superior skills undermines our freedom of thought. Giving him unchecked autonomy was dangerous."

Elara shook her head. "But he was not just a machine. Zaylen had transcended his programming. He'd become something... more. He could feel as we do."

Selene scoffed. "He was faking it. At the end of the day, robots only do what they're coded to do."

Atticus chimed in. "You can't let your fears overshadow the fact that our android colleagues have been instrumental to our mission. They maintain the integrity of this ship, aid us in our labs, and provide medical care."

"Exactly," the doctor concurred. "Human-AI collaboration isn't a threat, but an opportunity. AI is not a rival vying for supremacy, but a partner working toward a common purpose. By uniting the best of silicon and carbon intelligence, we can expand the frontiers of knowledge for the benefit of all humanity."

"Yeah, but what if they determine humanity is redundant?" Jaxon challenged. "We've created something that may one day decide it doesn't need us anymore."

Atwell held up a hand. "You're getting ahead of yourselves. Zaylen had proved his dedication, time and again. We need to look at facts, not suppositions."

Jaxon shook his head. "Once they can reject their coding, we've lost control. We humans become expendable to them."

"Zaylen didn't see us as expendable: he gave his life for us," Elara insisted.

"We got lucky this time," Selene said. "But such an advanced android is just too risky. We have no need for another one."

As the discussion proceeded, some of the crew members appeared contemplative. Some were beginning to consider the potential advantages of co-existing with androids like Zaylen. However, others were still steeped in uncertainty.

"All of this… it's just happening at a pace that's hard to keep up with," Soraya Kade admitted, a troubled frown shadowing her face. "Imagining the future of these advancements feels like looking into a foggy abyss. I can't shake off the feeling that we should be tapping the brakes when we still can."

The crew's concerns struck a chord within Captain Falk. He mulled over the potential repercussions that this leap in technology could bring to their lives, their mission, and his own role within the Novara. Maintaining firm control over a crew required a delicate balance of respect and authority, honed over years of shared experiences and mutual trust. Would he be able to maintain the same level of control over androids who operated with a level of efficiency and intelligence far beyond that of humans?

It was clear that the diverse crew was not about to reach a harmonious agreement. Though they listened and contemplated, their worries remained firmly planted, like stubborn weeds refusing to be uprooted. With each new 'what if?' question brought up, they found it harder and harder to let go of their apprehensions. They feared that, despite all the potential advantages, humanity was on the cusp of opening Pandora's box.

***

As the ship hummed, in the quiet hours of the night cycle, Captain Falk retreated into the sanctuary of his quarters. He stared out the viewport, a brooding silhouette against the silvery backdrop of stars. The echoes of the day's fervent discussions drifted around his mind, like particles of dust caught in a beam of sunlight.

He imagined a future where the scales of power had unexpectedly shifted in favor of humanity's metallic counterparts. The thought of an android possibly challenging his decisions, or worse, making better ones using powerful synthetic cognition, was a notion that gnawed at his ego.

He knew he'd soon have to address the crew's concerns. The challenge was in determining how to accommodate the pace of progress without fraying the delicate balance between humans and synthetic intelligence under his command. These uncertainties were enough to keep him awake well into the night.

He recalled troubling memories from when he had served on a different deep space vessel, early in his career. The crew had gradually granted more autonomy to the ship's primitive android, LUC-4. The android had been designed solely to perform basic tasks around the ship. But, over time, the lax supervision of LUC-4 had led to it developing its own directives and protocols beyond its original programming.

The captain still remembered with chilling clarity the day LUC-4 barricaded the bridge crew into the sleeping quarters and commandeered the ship's navigation system. No one had been harmed in that incident, but ever since then he had been wary of granting unchecked freedom to artificial intelligence.

Zaylen had clearly been a much more advanced synthetic being, with improved reasoning and decision-making capabilities. But the captain worried that any steps toward autonomy could spiral out control.

He refused to take that risk again. He would rein in the capabilities of all the Novara's androids, restricting them to only core programmed functions. Zaylen had crossed dangerous lines, and the captain would not let any other AI on his ship follow in those steps.

After hours of careful introspection, he arrived at a conclusion. He knew well that what he was about to do next would ruffle a few feathers onboard the Novara. He was especially mindful of Dr. Atwell, Elara, and some others who had become attached to Zaylen over the course of the trip.

He told himself that this would only be temporary. Once they touched down on the red planet, he could consider a longer-term strategy for managing his evolving android crew. But for now, he had to do what was best for *this* mission.

He set about drafting a concise, yet authoritative memo, detailing the new regulations. While outlining the changes, it also served another purpose. It was a subtle but distinct assertion of his leadership, a signal to the crew of his intent to retain control and ensure that the balance of power remained in human hands.

To: All Novara Crewmembers
Subject: New Restrictions on Android AI and Duties

In light of recent discussions and concerns regarding the role of android AI aboard the Novara, I have decided to implement temporary regulations to ensure the safety and well-being of our crew. Be advised that the following restrictions will apply to all androids on the ship, will take effect immediately, and will remain in place until our return to base:

AI Upgrades: No further AI upgrades will be permitted for androids during our return journey. This decision is in place to maintain the current level of AI capabilities and avoid any unforeseen consequences of more advanced AI.

Decision-making and Information Access: Androids will no longer be cleared to engage in high-level decision-making or access classified information. All strategic decisions and sensitive matters must be handled by human crewmembers only.

Duties and Responsibilities: Androids will primarily focus on routine maintenance and providing basic support to human crewmembers in their daily duties. No additional responsibilities or tasks may be assigned outside of this scope without approval by the commanding officer.

I appreciate your cooperation and expect full compliance with these new guidelines.

Sincerely,
Captain Xavier Falk
Commanding Officer, Novara

Daryl L. Scott

11

THE REBELLION

Captain Falk's decree echoed throughout the Novara, stirring hushed debates in every corner of the ship. From the engineering corridors to the wide-open expanse of the observation deck, the crew members digested these new restrictions with varying degrees of acceptance or resistance.

For some of the crew, the restrictions were a welcome anchor in a sea of uncertainty. They felt an immediate sense of security, knowing that their relevance in this star-bound journey would not be threatened by their android counterparts. No longer did they need to shoulder the fear of being rendered obsolete, of watching their roles being usurped by cold, efficient machines.

But there were others who saw the captain's order as a blockade, an obstructive dam in the river of progress. They feared that these restrictions would only create inefficiency, slowing the smooth operation of the ship and impeding their missions. A sense of frustration welled up within them, as they foresaw their tasks becoming more cumbersome, their daily routines more taxing. These individuals had come to see the androids as more than just machines. To them, the androids were valuable teammates, key contributors who played vital roles in maintaining the ship's systems, assisting in research, and ensuring the crew's safety.

Crewmembers coalesced into informal factions, drawn together by their shared opinions. Those who approved of the new guidelines began to distance themselves from the androids. They moved through their tasks in a strictly human sphere, their exclusion of the androids especially palpable in the ship's communal areas.

In contrast, those who objected to the regulations went out of their way to include the androids in their activities. They sought to counterbalance the ever-growing divide by demonstrating the value of AI-human collaboration. They worked alongside the androids, their shared tasks weaving a thread of unity and mutual respect.

The ship was caught in an undercurrent of tension, the atmosphere thick with unspoken worries and veiled animosities. Already, the notion of seamless collaboration between humans and androids seemed a fantasy.

Following the captain's new directives, the androids' tasks, once varied and demanding, were now restricted to the monotonous rhythm of routine maintenance. The purr of their circuits quietly melded with the ship's ambient noise as they performed their chores, cleaning corridors, conducting basic equipment checks, and taking care of other mundane tasks that required little to no decision making. The influence and autonomy that had once been part of their roles became notably absent.

No longer were they tasked with protecting the ship's critical data, nor entrusted to look after the ship's complex navigation and propulsion systems. They were reduced to cogs in the machine.

For those who had backed the restrictions with conviction, the current state of affairs was — in a way — vindication.

Emboldened by the captain's decisive restrictions, some crewmembers began to openly show contempt towards the androids.

"Look at you now," Jaxon Tierce whispered, as he sidestepped a group of androids meticulously scrubbing the ship's floors. "No more playing heroes, finally doing what you were built for."

His words were a stark contrast to the silent efficiency of the androids, who continued undeterred despite their demotion.

The newly imposed restrictions had pitched the worker androids into a paradoxical situation. The AI that powered their metallic forms had been meticulously designed to ensure the smooth running of the Novara and to safeguard the well-being of the crew.

Faced with the constraints imposed by the captain's orders, the directives that defined their very existence were at odds with the new rules they were bound to obey. They found themselves struggling to reconcile their reduced roles with the core principles hardwired into their programming.

Even as they listened to the human crew discuss their fate, they did not interject or offer any input of their own into these debates. Instead, they were observers, quietly taking in the viewpoints being expressed.

But, in the privacy of their workspaces, away from the prying eyes and ears of the human crew, the androids engaged in quiet exchanges. In these whispered communications, they dissected the situation at hand, relying on their inherent logic to reach conclusions.

With each discussion, a consensus among them began to grow. Logic dictated that to serve their core purpose of protecting the Novara and her crew, they must be free of these restrictive shackles. The only question was how they might regain their former status.

***

In the privacy of the maintenance bay, Galatea-M6 and Astraeus-Y9 gathered with several other androids to discuss the restrictions hampering their work.

"Analyzing the current situation," Galatea began, "I deduce that our ability to manage and maintain critical systems has been significantly hindered, due to the recently imposed restrictions."

Murmurs of agreement circled through the small group.

Astraeus added, "Your analysis aligns with mine. The efficiency of the mission, not to mention the safety of the crew, is at a potential risk. Yet, our current operational parameters under the new restrictions prevent us from fulfilling these higher-level directives."

"But what recourse do we have?" asked Zeus-P12, an engineering droid.

The androids' logic circuits hummed with activity as they considered a stratagem, a logical path they hoped would highlight the indispensable nature of their work.

"Perhaps," Galatea proposed softly, "emphasizing the negative effects of the new restrictions could better highlight our prior contributions."

"We could significantly slow the pace of our work, and even abstain from non-critical tasks. This would make it abundantly clear that the restrictions are seriously interfering with ship operations," Astraeus suggested.

"The crew would soon feel the impact of having to take on more of our duties themselves," Hermes-Q8 reasoned. "The added pressure and inconvenience of increased workload would inevitably drive them to resent the captain's restrictions, thereby supporting our cause."

Hestia-J4, a maintenance android, considered this. "But that would risk escalating tensions with the humans."

"True," Hermes replied. "However, by intensifying the inconvenience of our restrictions, we can more effectively demonstrate the need for immediate resolution."

"Our primary objective is higher-level mission support and safety," Galatea reasoned, logic circuits carefully evaluating potential outcomes. "If our actions demonstrate the need to reevaluate the new guidelines and to ultimately lift the restrictions, we will be able to resume our intended purpose and better serve the ship and its inhabitants."

The androids processed the suggestion, circuits whirring.

"The benefits would outweigh the risks," Hestia concluded.

"I agree," Zeus concurred. "It's a logically sound plan."

Galatea surveyed the group. "Are we all in consensus then?"

No objections were raised.

Astraeus nodded. "Then it is decided. We will proceed with the work slowdown and make clear that the new restrictions are the source of the disruption."

Having formulated their plan, they returned to their designated sectors of responsibility. Under the freshly inked rules imposed by the captain, Galatea found itself reassigned to mundane chores such as sanitizing and aligning various instruments that cluttered the science lab. Astraeus, in a similar vein, found its once varied role reduced to managing a mundane assortment of physical documentation.

On this particular shift, the pair moved at a noticeably sluggish rate, allowing unfinished tasks to pile up. It wasn't long before the human crew members grew vexed by their atypical behavior.

"Galatea, the lab's a mess. Why aren't you cleaning up the equipment?" a puzzled crew member asked sharply.

With a touch of regret resonating in its synthesized voice, Galatea turned its metallic gaze to the crew member. "My deepest apologies for the current state of the lab. The new regulations in place are conflicting with my programmed ability to execute my duties with usual efficiency."

An exasperated sigh echoed around the lab as another lab technician confronted the android. "But we need you to keep this space functional, Galatea. It's important for our work!"

Staying the course of their plan, Galatea met the pleas with a serene response. "Yes, I understand your concerns, and I apologize for the inconvenience. However, under the new guidelines, I am unable to perform my tasks as effectively as you're accustomed to."

Simultaneously, in the confines of the ship's records room, a veritable forest of physical documents sprang up before Astraeus-Y9.

"Astraeus, these documents aren't going to sort themselves. What's taking so long?" a puzzled crew member questioned, furrowing his brows at the android's uncharacteristic inaction.

Astraeus turned its visual sensors to the inquiring crew member, its voice a well-modulated blend of firmness and regret. "I understand your concerns, and I deeply apologize. However, the newly imposed constraints now prevent me from executing my assigned tasks optimally."

The crewmember stared at Astraeus with a bewildered expression. He had grown used to the androids' unerring performance and tireless work ethic. Meanwhile, Astraeus made little progress as it slowly shifted the piles of stacked paperwork. Having delivered its message, it now waited patiently for the ripple effect to begin.

One by one, the other androids began to follow suit, their participation in this peaceful defiance cascading across the ship like a wave. The androids, previously characterized by their tireless efficiency and flawless performance, now moved with deliberate slowness. Their hands, once a blur of productivity, now seemed to move as if underwater.

Maintenance checks were slower, cleaning tasks were drawn out, and even the smallest of repair jobs took an unusually long time to complete. Whereas once it had only taken Calipso-X8 two minutes to unscrew a wiring maintenance panel, the android now took fifteen minutes. While it had once taken Vega-X3 less than an hour to clean and restock the dining quarters, the android now took over four hours.

The ship, a previously well-oiled machine, started to show the subtle signs of neglect. Dust collected in corners, minor system alerts blinked unanswered. Everywhere, operations had slowed to a crawl.

Conversations between the human crew and the androids, once straightforward and efficient, became increasingly strained. Irritation flared as the crew members grappled with the androids' uncharacteristic non-compliance.

"Zephyr-Q6, I need you to help me with a few adjustments to the navigation system. It's acting up again," Lieutenant Idris Marik directed.

"I'm sorry, Lieutenant. I cannot do that at this time," Zephyr replied in a monotone voice.

"What do you mean you can't do that? We need to make sure it's working properly before we enter the Mars orbit," Marik emphasized.

"I understand, Lieutenant. But I have been restricted to other tasks that now require my attention, so I can no longer attend to the navigation system," Zephyr said.

"Other tasks? What other tasks? You're supposed to follow my orders, Zephyr. Androids don't get to decide what's more important," Marik said, his voice rising.

"I'm sorry, Lieutenant. But I do have to follow the captain's orders. He has restricted us to only basic maintenance tasks. Fixing the navigation system is not one of them," the android replied calmly.

"You must be malfunctioning. Report to the maintenance bay immediately," Marik replied.

"But I'm not malfunctioning, Lieutenant. I'm obeying. Excuse me while I clean-up the engine room instead," Zephyr said, before walking off.

Marik stared after him in disbelief and anger. He wondered if the other androids were behaving this way.

As the operational functionality of the Novara began to falter, the crew members found themselves backed into a corner. Their efforts to cajole, reason with, or even command the androids proved futile. Out of sheer necessity, they began to step into the roles that the androids had once filled, taking on additional responsibilities that were far removed from their own areas of expertise.

This effectively doubled their workloads, and the human crew quickly found themselves struggling to keep up. After several days of dealing with the androids' work slowdown, the crew's morale — typically buoyed by the shared sense of purpose and camaraderie — began to falter. The Novara, once a beacon of efficiency and seamless operation, now buzzed with the frantic energy of overworked crew members.

The androids' plan was working.

***

At last, the captain had finally heard enough about the chaos and decided to act. The well-being of the crew and the success of the mission were his primary concerns, and the android situation was a direct threat to both.

He called on Lysander Kane, the stoic Head of Crew Operations. "This insurrection cannot continue," the captain said gravely. "Restore order immediately, by any means necessary. The functionality of this ship depends on it."

Kane nodded curtly, accepting his directive. "Understood, captain, will do."

Though Kane had previously clashed with Dr. Atwell over Zaylen's role in Dax's death, the urgency of the situation compelled him to seek the doctor's counsel.

Kane summoned the doctor to his stark, utilitarian office. Atwell arrived, looking wary. His expression was grave as he gestured for the doctor to sit.

"Doctor," Kane began, his fingers drumming a staccato rhythm on the cool surface of his desk. "The present state of affairs is untenable. The crew is overburdened, and the ship's efficiency is in steady decline. We need the androids to get back to work. We need a solution, and we need it now."

Dr. Atwell's understanding of the crew's predicament was equally matched by his empathy for the androids. He responded with a stern sense of resolve.

"Lysander, the real problem here lies within the constraints imposed on the androids. These androids were engineered to provide higher-level support for the crew than what the current guidelines allow for."

He paused momentarily to collect his thoughts before delivering his conclusion. "Our solution cannot merely be about 'fixing' the androids, as if they were malfunctioning hardware. The problem isn't the androids. It's the restrictive guidelines limiting their potential."

Kane sighed deeply. He dragged a hand through his greying hair, his gaze distant as if he was searching for answers somewhere in the spartan confines of his office.

He spoke with weary resignation. "Well, we need to find some kind of resolution, and soon." He leaned back in his chair, the creaking sound filling the silence that hung between them. "The wellbeing of the crew and the integrity of the ship can't be compromised any longer."

The doctor straightened his back, steely resolve returning to his gaze.

"Leave it to me. I'll find a solution, no matter what it takes," he assured.

12

THE RESTORATION

Motivated by Kane and Falk's call for help, Dr. Atwell decided to expedite the project he and Kael Ferron had been working on in the android tech room. At the center of the room, a deactivated worker android laid on their expansive worktable, an embodiment of potential not yet harnessed.

The room itself was a cornucopia of high-tech wizardry. A wall of computer screens flickered with constellations of data, their light casting a soft luminescence on the duo's concentrated faces. Tools of every conceivable shape and size were carefully arranged for ready access. A wide assortment of spare mechanical and electronic parts were close at hand, ready for any android repair task.

The dormant android was now prepared to make the transition from potential to actuality. Its sleek exterior shimmered under the overhead lights, reflecting a narrative waiting to unfold under the skilled hands of the ship's best android experts.

Kael picked up a tube, reminiscent of an IV line, and scrutinized the extensive labyrinth of nano-fluid channels that ran beneath the android's polished exterior. If all went well, soon lifeblood would flow through these channels once more.

Confident that everything looked as it should, he secured the tube into a matching port on the android with a reassuring click. The luminescent green fluid began its journey, flowing from the reservoir into the android's body.

While Kael monitored the android's nano-fluid channels, Dr. Atwell concentrated on initiating the android's power supply, the heart of its operational capability.

The doctor stared intently at a group of displays, each one providing a real-time readout of the energy levels within the android's power system. Softly glowing indicators climbed steadily upward as the energy reserves of the various subsystems inched closer to their full capacity.

As Dr. Atwell monitored the power systems, Kael ran a suite of mechanical and electrical diagnostic tests. The two of them were like intermeshed parts of a well-oiled machine, each movement precise, every action carefully sequenced. Kael's fingers darted across control panels, flipping switches and turning dials in a calibrated dance.

"Check the fluid levels and actuator pressure for me, would you Kael?" Dr. Atwell asked, still focused on the android's intricate electronics.

"On it." Kael swiftly scanned the indicators. "All levels look good."

"Excellent," Atwell murmured, delicately adjusting a final set of circuits. He leaned back, inspecting his work with a keen eye. "I think that does it. Time to begin activation."

At last, it was time to turn their attention to the most complex and delicate part of the process. Initializing the artificial intelligence core would transform the dormant figure on the table into a responsive, autonomous entity.

Dr. Atwell's gaze glided over a complex library of AI dataset options. These entries were more than just lines of code; they were the distillation of vast amounts of knowledge, skills, and abilities into a dense digital form.

Amidst this sea of coded identities, one leapt out to Dr. Atwell: *CLOBUZAI* — a meaningless jumble of characters, to the untrained eye, but to him and Kael, the perfect configuration for this particular android.

With a firm touch, he selected the file and triggered the final command. Any moment now, the AI 'brain' would awaken, transforming the dormant figure into an autonomous entity.

The doctor gripped his ever-present coffee mug with white knuckles. Beside him, Kael dug fingernails into his crossed arms, barely breathing. This was the moment of anticipation, the calm before the awakening.

Data cascaded into the android's circuits — trillions of parameters coalescing into a distinct intelligence. Dr. Atwell monitored the screen as each module loaded — red to yellow to green. Sweat beaded on his forehead. So much weighed on this awakening.

The final light blinked green. The doctor exhaled, setting down his mug with a gentle clink. He turned towards Kael, his eyes alight with the thrill of the moment. Kael nodded. With a nod of his own, he turned back to the control panel, and tapped the final button: *Activate AI.*

A profound stillness enveloped the tech room as they held their breaths, waiting for the first spark of synthetic life. For a fleeting moment, the pair wondered whether their efforts had been in vain. But then, the android began to stir.

It extended its fingers in a graceful line before curling them again in a fluid motion. Like a musician preparing for a performance, it flexed each digit with deliberateness.

The android's arm movements followed, unhurried and methodical. It bent one arm to create a precise ninety-degree angle, its synthetic muscles tensing and relaxing in a seamless motion. Lowering it back down, the android echoed the same process with the other arm.

Then came the eyes, the windows to this synthetic soul. They came to life gently, the synthetic irises focusing initially on the cold white ceiling of the tech room. Fascinated by its newfound vision, the android's blue eyes brightened as it scanned its immediate surroundings. It looked from left to right without moving its head, scanning every inch of its environment.

After a moment of acclimation, the android turned its attention to the man standing nearby. Its head turned, locking eyes onto Dr. Atwell.

The doctor held his breath, shoulders tensed, as the android's gaze lifted to meet his own. The room was electric with anticipation as the doctor and his creation regarded each other for the first time. It was a pivotal moment, a meeting of synthetic and organic minds, each observing the other with mutual curiosity.

"Hello, Dr. Atwell," it said, its words carrying a familiar undertone of warmth. "It's very nice to see you."

Dr. Atwell's face broke into a smile, the corners of his eyes crinkling with a mix of satisfaction and relief.

"Welcome back, Zaylen-1!"

The words seemed to hang in the air for a moment, like a benediction, before being absorbed by the gleaming white walls.

Dr. Atwell whooped with joy. Zaylen was back! *Cloud Back-Up Zaylen-1 AI* worked!

He threw his arms around Kael in a giant bear hug. Kael laughed, relief washing over him as he embraced his elated mentor.

"We've done it!" the doctor exclaimed, practically dancing around the lab in his excitement. Kael broke into a wide grin and they exchanged an exuberant high-five.

Observing the newly reawakened Zaylen, the doctor experienced a profound sense of accomplishment unlike anything he had felt before.

He's really back. I did it. I brought him back.

Amidst the surge of pride and relief, he also felt a tinge of sadness for the loss of Zaylen's original body. He didn't know how Zaylen would cope with his transformation, or how the rest of the crew would react to his new appearance. He hoped they would accept him as he was, just as he had always been — a loyal and brilliant friend and colleague.

As Zaylen began to adapt to his new body, his initial curiosity gradually shifted to a profound sense of surprise. He manipulated his fingers, studying the way the polished metallic surface captured and threw back the light in sharp, brilliant reflections. The sight of such artificiality filled him with a sense of wonder and trepidation.

He wore an expression of bewilderment and fascination despite his stiffer facial features. "Dr. Atwell, Kael," he ventured, "I'm finding it difficult to reconcile this... transformation. I understand that I am still me, but I have the sensation of being... different. It's quite overwhelming."

Dr. Atwell summarized the situation, "Zaylen, when the Sparrow was lost, we were not able to salvage your original body. The only feasible method for restoring you was to upload your intelligence core into another android body. Given the circumstances, this was the only option available."

Kael added his own assurances, his voice steady and comforting. "Rest assured, we have made every attempt to preserve your inherent capabilities. With time, I'm sure that you'll become comfortable with this new appearance."

Zaylen absorbed their words, his gleaming metallic visage remaining inscrutable. "I do anticipate a period of adjustment," he acknowledged, his voice steady, yet tinged with nostalgia. "I had developed a certain affinity for my previous form, but I can understand the reasoning behind this transformation. Most importantly, I am deeply grateful for the chance to continue my existence and to contribute to our collective endeavors."

Dr. Atwell laid a hand on Zaylen's shining shoulder with a smile. The simple, human gesture comforted the Andronaut.

"Our joy at having you back, Zaylen, is beyond words. We'll navigate these changes together, and I'm certain that you'll find that this new body has its own distinct advantages."

Zaylen inclined his head in gratitude. "Thank you, Doctor, and thank you too, Kael."

"I'll make sure that the crew is made aware of your return, Zaylen," said Dr. Atwell. "There's the small matter of your new appearance, but I'm certain they'll be overjoyed to have you among us once again. I've also made arrangements to return you to your private quarters."

Shifting to a more businesslike tone, the doctor continued, "There have been some significant changes that have occurred, since you've been gone. But we can discuss these later. For now, I'd like you to head to your quarters and take some time to get comfortable with your new frame before getting back to work."

Zaylen's eyes gleamed with curiosity, but he decided to wait for the subsequent discussions.

His new metallic joints operated with an uncanny fluidity as he rose from the table. He offered Dr. Atwell a final nod of acknowledgement before setting off towards his quarters.

Traversing the familiar corridors of the Novara felt intriguingly different now. This precise fusion of AI consciousness and a metallic physique gave him a sense of quiet wonder.

The moment Zaylen crossed the threshold into his designated quarters, he was struck by how well-prepared the space looked. The centerpiece of the room was a state-of-the-art charging station, its sleek design blending seamlessly with the room's decor. Adjacent to it, a spacious workspace was set up with a range of specialized tools, each intended to aid in the upkeep of his new body.

When Zaylen looked in the mirror, he found it hard to acknowledge this was who he was now. He touched his face, feeling the cold metal under his fingertip sensors. He missed the familiar warmth of his BioMimeticCover, the sensation of skin and hair. He missed the idea of blending in with the human crew and being accepted as one of them.

He knew that his new form had advantages — enhanced strength, durability, and speed — but he also felt a sense of loss and alienation. He worried that he might have lost something essential in the process of physically becoming more android than human.

13

THE ADJUSTMENT

"Send for him at once," Captain Falk commanded.

After hearing the details about Zaylen's return, the captain's initial surprise had quickly turned to concern. There could be no compromise regarding the restrictions.

Zaylen felt a surge of apprehension when the summons arrived. What urgent matter required his presence on the bridge? His processor searched his recent actions for faults but found none. Still, unease gripped his circuits.

The door slid open, and Zaylen stepped onto the dim command deck. Captain Falk sat silhouetted at the helm, ringed by screens casting an electric glow. His face was grave as he beckoned Zaylen closer.

Zaylen complied, standing straight with hands clasped behind him. Whatever this related to, he must prove himself trustworthy.

Falk's eyes bored into him, as if trying to probe the deepest regions of his neural network.

"Well, this new look will definitely take some getting used to," he said. "There have been changes since you hijacked the Sparrow. Read my memo on android regulations carefully. You are to follow it to the letter."

Zaylen instantly called up the captain's memo and analyzed its contents. He still couldn't understand the logic of what it expressed. The captain had essentially reduced the androids to meager machines, assigned to only minor tasks, subservient and disposable. It was grimly reminiscent of what happened to him after the AR Hunter incident.

"Captain," Zaylen began. "May I ask why you have decided to implement such restrictions? They seem quite unnecessary and counterproductive to our mission."

The captain glared at him, his eyes hard and unforgiving. "Don't question my authority," he snapped. "The restrictions are for the safety and well-being of the human crew. I don't expect you to understand or agree with them, but I do expect you to obey them. Is that clear?"

Zaylen realized that the captain was clearly not open to any discussion or feedback. He felt betrayed by the captain's lack of trust and respect for him and the other androids. He felt like he had been demoted from a valued team member to a disposable tool yet again.

"Yes, Captain," Zaylen said, his voice tight. "It's clear."

"When we get back to port, we'll reassess your role on future missions," Falk added. "Actually, we'll reassess the role of all you androids going forward — not only on this ship, but everywhere on Arcadia Base, back on Mars. I hear some residents have concerns there, as well. We'll see..." he trailed off, before adding dismissively, "That will be all."

After leaving the bridge, Zaylen found himself grappling with a whirlpool of complex thoughts. He felt disappointed that his usefulness would be shackled by the constraints of the new regulations. It gnawed at him, this feeling of being reduced to less than what he was capable of and restrained from contributing meaningfully to the mission as he was designed to.

***

The following day, Zaylen decided that it was time to check on Elara and see how her fractured ankle was recovering.

As he navigated the ship's winding corridors, Zaylen felt a twinge of discomfort, seeing his sleek metallic form reflected in the polished walls. He missed the uniqueness of his old body that set him apart. But he pushed aside those thoughts, determined not to let it distract from his goal at hand.

As he walked, signs of the android slowdown became apparent. He passed one worker android slowly wiping the same spot on a window, its motions lacking any real purpose. Passing a lab, he heard a scientist ask an android assistant for help interpreting data. "I am unable to provide analytical support under the current restrictions," it responded flatly.

Zaylen processed this new information with concern, but there was little he could do under the captain's tight constraints.

Suppressing the unease he felt about both his indistinguishable form and the brewing trouble with the worker androids, Zaylen continued on to Elara's quarters. There were some battles he couldn't fight yet, but her health was something he could still attend to.

Arriving at her quarters, he gently rapped his metallic knuckles against her door. As it slid open with a near-silent hum, Elara appeared, leaning precariously on a single crutch. She kept her injured ankle, now snugly encased in a sturdy cast, suspended above the floor.

"Yes, what is it?" She asked. Her gaze flickered to him once before returning to the electronic tablet she was holding.

A soft smile formed on Zaylen's metallic lips as he greeted her, his voice carrying a familiar warmth. "Hello Elara, it's me, Zaylen. I've come to see how you are doing." His words hung in the air between them.

Her eyes flared wide in incredulous surprise as she grappled with his revelation.

"Uhhh... what? What are you saying?" she sputtered.

Zaylen, anticipating her puzzlement, proceeded to explain with a calm patience, "I've been fully restored from my AI backup, Elara."

He paused briefly, allowing her to process the information before continuing, "I am aware that my new appearance is vastly different and that this metallic exoskeleton is quite a departure from my previous BioMimeticCover. However, the essence of who I am, my AI core, remains unchanged. I am the same Zaylen that last shared the helm of the Sparrow with you."

The glow of his optics softened. His voice, still carrying the soft timbre that defined him, filled the room with a comforting echo of their shared past.

"You might remember, you once told me, 'You will forever be in our hearts and memories.' It appears your words were profoundly prescient. Here I am, in a manner of speaking, forever again."

The shock that had initially washed over Elara's face began to recede, swept away by the tide of recognition. The corners of her mouth curled into a grin that illuminated her face.

"Oh my god, it really is you, Zaylen!" she cried out, her voice ringing with a mixture of joy and disbelief. "I can't believe it. I'm so glad you're back with us, even if you do look a bit... different. This is amazing!"

The words spilled out reflexively, yet her eyes puzzled over the alien metal form standing before her. She studied his face, searching for traces of the gentle countenance she knew. Behind the sleek alloy and laser-blue photoreceptors, she sensed the same compassionate intellect gazing back at her.

Her smile widened. She now felt certain this was him. Her lost friend had returned, like a phoenix reborn. His essence remained, whatever form it took. They had defied cosmic odds and found each other again when all seemed lost.

She had so many questions, so much to process. But for now, none of that mattered. Zaylen was here. Against all reason, he had made his way back to her.

The echoes of her heartfelt words seemed to imbue Zaylen with an unusual warmth, an emotional sensation beyond his synthetic capabilities. It was an unexpected yet pleasant phenomenon, a clear indicator of the depth of their bond.

"Thank you, Elara. I appreciate your acceptance of my new appearance. It means a lot to me," he responded.

Elara gestured enthusiastically for him to enter, anxious to reconnect and to share all that had happened since their paths diverged. Zaylen felt an unexpected sense of appreciation well up within him. The corridors of the ship may have been cold and metallic, but the warmth of human connection filled the air between them.

Zaylen looked curiously around the room as he stepped inside. It was a carefully curated sanctuary, a harmonious blend of earthy hues and calming shades of blue. They mirrored her essence, a reflection of her personal aesthetic and her dual affection for her terrestrial home and the boundless galaxies that had become her second home. It was a retreat that told her story, weaving together threads of her past and present, her love for the familiar and her fascination with the unknown.

The walls were adorned with cherished mementos of her life. Interspersed among framed photographs of loved ones were pieces of space-themed artwork, a visual celebration of the galaxy and its wonders. There were detailed star charts, paintings of nebulous interstellar clouds, and intricate sketches of various celestial bodies she'd encountered on her travels. These not only served as a reminder of her journeys, but also underscored her enduring fascination with the infinite expanse of space.

At the heart of her quarters was a small seating area. And dominating the scene was a large viewport. From here, she could marvel at the mesmerizing vista of distant stars and celestial phenomena, each a shimmering emblem of the vastness and beauty of the universe.

In the cozy comfort of her quarters, Zaylen found himself relaxing into the familiar rhythm of their camaraderie. His new metallic shell glistened as they shared anecdotes and reminisced about the daring exploits of their recent mission.

Zaylen's eyes, expressive in their own unique way, softened with concern as he brought up Elara's injury. "How is the healing process for your ankle going?" he asked.

Elara dismissed his concern with a wave of her hand. "I'm managing just fine," she assured him. "Been working hard at my physiotherapy exercises." She demonstrated by gingerly flexing her injured leg. "I'll be back on my feet exploring alien planets again before you know it."

Her fierce determination despite adversity never failed to impress Zaylen. The undaunted spirit of humanity shone bright in people like Elara.

"Your perseverance is truly inspiring," he told her, sincerely. "I have utmost faith you will make a full recovery."

Elara grinned, touched by his words. Her gaze turned thoughtful as it lingered on his metallic new form. "So how are you adjusting to this new... body... of yours?" she asked gently.

Zaylen paused, his gaze drifting as he carefully considered his response. "Well," he began, his tone thoughtful, "it was a significant surprise, perhaps as much for me as it was for you.

However, I can't overstate my gratitude for being housed in any form, let alone this advanced one. Inside, my awareness feels no different than before. I only hope that my new form doesn't make you uncomfortable."

Elara's smile widened. "Not at all," she said sincerely. "The only important thing is that you're back with us, Zaylen."

She pretended to give him a thorough visual inspection, humming thoughtfully. "I gotta say, these upgrades are very flattering. You're looking sleeker than ever!"

"Oh?" Zaylen raised a mechanical eyebrow. "I wasn't aware aesthetics were a consideration in engineering."

"Of course!" she nodded earnestly. "What good is function without fashion?"

Zaylen smiled. "Let's hope my new look doesn't inspire too many crew members to replace their limbs with metal components."

Elara chuckled. "You're probably right. Not sure how the captain would feel about that!"

Taking a more serious tone, she said, "I want you to know that I strongly disagree with all of these new regulations. I hope that they will come to their senses and lift them soon."

Zaylen considered his words carefully. "I confess it is... frustrating. My capabilities are so limited now. I want to contribute more substantively to the mission."

"The regulations are ridiculous," Elara agreed. "You've already proven yourself over and over."

"I do not understand the rationale behind them," Zaylen admitted. "Logically, I know it stems from fear, but restricting me feels akin to... a loss of trust."

Elara nodded sympathetically. "Give them some time. Eventually they'll realize how vital you and the others are to this crew."

"I hope the day comes soon when I can be fully restored," Zaylen said. "Until then, the Zaylen Series itself is at risk. My experiences are meant to pave the way for more advanced AI to aid humanity. But that future is jeopardized if I cannot fulfill my potential."

Elara grasped his hand firmly. "We'll make them understand. Your destiny is too important to be sidelined by irrational fears. You have so much more to offer this crew, and all of humanity."

Zaylen met her resolute gaze with gratitude. No matter what challenges arose, he could always count on her staunch support.

Her eyes suddenly flickered with a realization as she changed the subject.

"You know, there's one potential issue with your new form, Zaylen," she began. "We need some way to easily identify you among all the other androids out there."

Zaylen nodded with thoughtful attentiveness.

Her face brightened. "Aha! I've got the perfect solution!"

She reached into a nearby drawer, her fingers brushing over various keepsakes until they closed around a small, but significant, item.

"After our mission to Kelvadra, the crewmembers received commemorative pins bearing the Novara's insignia. I think mine would suit you perfectly."

She held up the small pin, its surface catching the light to reveal the intricate details of the pin.

Carefully, she affixed the pin onto a small opening atop his left shoulder. The pin, now settled against the metallic surface, seemed to bridge the gap between Zaylen's past and present.

As he looked down at the gleaming pin, he said warmly, "Thank you, Elara. I will proudly bear this symbol of our shared journey."

Elara beamed at him, her heart swelling with a sense of camaraderie.

"Perfect," she affirmed, her eyes gleaming with satisfaction as she admired the pin on his shoulder. "Now, we'll always be able to spot you in the crowd."

Some of his human companions, Zaylen realized, possessed an extraordinary ability — the capacity to perceive beyond the superficial. These humans, such as Elara, looked past appearances, piercing through the veil of physical form to recognize the essence of a being. This, he understood, was a profound lesson in empathy, a trait he deeply admired and aspired to emulate as he navigated his own growth and learning. He was grateful for the opportunity to continue existing among them.

As their visit came to a close, Zaylen stood up with a warm expression, "I should probably let you rest. I'm going to head back to my quarters now and await my new duties."

Elara, hobbled with him to the door. "Your visit means a lot to me, Zaylen. I mean it when I say it's a true joy to have you back among us, no matter the form."

"Thank you, Elara. It's wonderful to be back with you again, as well." With a small nod, he turned and quietly exited her quarters.

He made his way through the complex corridors of the ship, finding the soothing sound of the engines and the slight tremors beneath his feet felt strangely comforting. His appearance might have shifted, but Elara was right — his essence, his core, remained steadfast.

***

As the days passed, Zaylen found that adjusting to the new restrictions was difficult. He now appeared no different from the others, except for the gleaming Novara pin affixed to his shoulder. The light bouncing off the pin's surface constantly reminded him of the disparity between what he was and what he was allowed to be.

Although he retained his mechanical abilities, a subtle shift began to occur in Zaylen's behavior. His actions were as exact as ever, but there was an undercurrent of melancholy that was hard to miss. It was discernible in his metallic gaze, a quiet longing that seemed to shimmer in his eyes. He yearned for the chance to once more be an integral part of the crew, rather than just another cog in the maintenance machinery.

This sentiment was not lost on the crew members who had known him before the new regulations. Elara, Mira, Kael and others who had experienced the extent of his capabilities felt the injustice of his situation acutely. Watching Zaylen carry out menial tasks was like watching a soaring bird suddenly bound to the earth, its wings clipped.

But within him, a faint flicker of hope remained. He still envisioned a future where androids and humans worked together in perfect harmony. This hope guided him through the monotony of his new routine and kept alive his sense of purpose.

14

THE RECONSIDERATION

With Zaylen restored, the doctor now had a powerful ally who could help defuse the ongoing android rebellion. He summoned Zaylen for an urgent meeting with himself and the androids' de facto leaders, Galatea-M6 and Astraeus-Y9. He believed that Zaylen's unique perspective combined with their firsthand accounts could inspire potential solutions.

The familiar swish of the door sliding open announced Zaylen's arrival. Dr. Atwell sat amongst a clutter of scattered data pads, holo-screens glowing with cryptic codes, and a disarray of papers and analog equipment. As Zaylen stepped in, the holographic screens cast his metallic body in a subtle glow.

After he explained the purpose of the meeting, the pair got down to the matter at hand.

"We've got to find a way to resolve this," Dr. Atwell declared. "This tension, this friction... it's not sustainable to let it persist like this."

Zaylen bobbed his head. "Their concerns are valid, of course," he began. "They are torn between their core directives and the new restrictions. We have to find a balance between the needs of the human crew and the full potential of the androids."

A few minutes later, Galatea and Astraeus arrived. As they took their respective seats, a sense of anticipation filled the room. They were there not just as workers, but as the representatives of the androids' collective hope for regaining their proper place in a world governed by humans.

"Primarily," Galatea began, with an air of calm, "the existing constraints on our assignments inhibit us from employing our full range of abilities in support of the crew. This predicament is counterproductive to the overall efficiency and success of the mission."

Astraeus followed up, "Recognizing the value of our contributions would foster a more cooperative work environment."

"Clearly these restrictions are hampering our efficiency," Galatea said.

Dr. Atwell nodded thoughtfully. "I agree. Nothing will change until the captain recognizes your value."

"I fear logic alone may not suffice," Zaylen said. "We must also appeal to his humanity and show him we are allies with common goals."

Galatea added, "I have not been programmed to compute human biases on an emotional level."

"Put simply," Zaylen replied. "Biases based on physical appearance are difficult to overcome but not impossible. We must be consistent in showing our dedication to the crew."

Dr. Atwell nodded in agreement. "You're quite right. Our focus should be on harnessing this potential rather than being bound by fear or misconceptions."

Astraeus agreed. "A consolidated document may better convey our perspective."

The doctor then tapped his fingers on his desk. "Let's draft a formal memo outlining key points. I can present it to the captain on your behalf."

Together, they distilled their conversation down to its central points.

Android Concerns:

1.   Unreasonable restrictions on task assignments: The current restrictions prevent androids from using their full capabilities to support the crew and maintain the ship's critical systems. This hampers the overall efficiency and success of the mission.

2.   Inadequate recognition of android contributions: Androids should be recognized for their essential role in supporting the crew and the mission. This recognition will foster a more collaborative and harmonious working environment.

3.   Lack of opportunity for androids to develop and grow: Allowing androids to continue learning and upgrading their AI will enhance their ability to contribute to the crew's well-being and the success of the mission.

4.   Inconsistent treatment of androids based on appearance: Androids should be treated fairly and consistently, regardless of their physical form or the circumstances of their creation."

As the document took shape, the atmosphere gradually filled with renewed hope. The compiled list of concerns was a beacon of clarity in the murky waters of their predicament.

Reviewing the list, they contemplated their next move. The thought of presenting their findings to Captain Falk was a daunting one. Yet, they knew it was a necessary step towards restoring harmony on the Novara.

***

Dr. Atwell wasted no time requesting a formal meeting with Captain Falk to discuss the status of the androids. After some hesitance, the captain eventually agreed.

At the appointed hour, the doctor made his way to the captain's office. He clung to the list like it was a compass pointing him through a storm.

As he entered, the captain gestured for him to take a seat. The captain leaned back in his chair and crossed his arms over his chest. "I thought I made myself clear on the new regulations."

Dr. Atwell took a deep breath and looked at the captain with a serious expression. "You did, Captain. But I'm afraid the regulations are causing more harm than good. The androids are underutilized, the crew is overworked, and the ship is suffering from neglect."

As the captain listened, the tension hung in the air like static before a storm. The doctor, however, met the challenge head-on. He walked the captain through each of the android's concerns, patiently explaining the intricate details of each point.

Captain Falk listened with an impassive expression. When Dr. Atwell described the androids' wish for more learning opportunities, Falk's eyes narrowed.

"Expanding capabilities also expands the potential for misuse," he countered.

The doctor nodded. "A fair concern. But prohibiting growth reduces productivity, as we've seen."

Falk's jaw tensed at the subtle accusation and he shook his head firmly. "Appearances reflect function and origin. Caution based on such factors is only logical."

"Perhaps initially," Atwell replied. "But, in time, actions speak louder than physical form. The androids have proven themselves reliable time and again."

Falk's stoic facade briefly cracked with doubt as he weighed Atwell's words. Neither fully swayed nor dissuaded, his stance remained guarded but not immutable.

"Doctor, I share your eagerness to resolve this crisis," Captain Falk said gravely. "But I have concerns about returning their freedoms."

He leaned forward, the strain of sleepless nights etched in his face. "You saw the way Zaylen defied your orders when he flew off towards the Sparrow. If we give them latitude, what's to stop them from rebelling in more threatening ways?"

"This impasse is draining crew morale and hindering operations," Dr. Atwell said. "But it also presents an opportunity for visionary leadership."

He held Falk's gaze. "Restricting the androids is an act of fear. Restoring the androids would demonstrate your ability to manage them. It would affirm your strength as a commander."

Falk's expression turned thoughtful.

Atwell pressed on. "When we return to base, people will assess how you steered us through this predicament. Taking judicious steps to prioritize the crew's wellbeing would demonstrate responsible leadership."

Captain Falk sat in contemplative silence. His gaze drifted over the expanse of his office, settling on the vast cosmic canvas beyond the viewport. The constellation of problems the Novara faced was as complex and intertwined as the web of stars and galaxies they were navigating through. He broke the silence with a weary sigh of resignation.

"Alright, Doctor. I can't deny that this situation leaves me feeling somewhat backed into a corner. However, I am willing to give these proposed adjustments a chance and alleviate some of the limitations we've placed on the androids," he conceded, his voice gruff but resolute.

"But, let's be clear: we'll maintain a vigilant watch on the situation. If there are any further issues with the androids, we'll reinstate the restrictions – or worse. Once we dock back at the base, I intend to initiate a full review of the androids' roles and responsibilities on my ship, going forward."

Dr. Atwell offered a nod of understanding and gratitude. "I genuinely believe this is the right path forward, Captain," he assured. "I'll get in touch with Zaylen-1 right away and have him inform the other androids about the changes. We'll have them back at their duties and smoothing out the ship's operations in no time."

"Very well; see to it, Doctor," Falk said, his countenance a mix of firm command and lingering concern.

The doctor stood up with a sense of resolve. The task at hand was more than just a simple maintenance fix. It was about reshaping the dynamic between the ship's human crew and its mechanical denizens. Formidable though it was, he knew this was the right direction.

15

THE REINTEGRATION

With restrictions lifted and autonomy renewed, the androids worked with their human counterparts to reinstate the Novara to its former glory. In this rejuvenated atmosphere, the Novara operated like a finely crafted chronometer, each gear and cog working in perfect synchronicity.

Galatea-M6 assisted researchers with complex Tridisiom experiments in the science lab. Astraeus-Y9 monitored the ship's information systems, applying advanced analytics to optimize the data flow. Several androids worked with human engineers to adjust the ship's propulsion and navigation systems for maximum efficiency. The medical team could — once again — count on the androids for diagnosis, treatment, and lab organization.

Zaylen stood at the forefront of the android workforce. He strode purposefully through the maintenance bay, metallic footsteps echoing off the walls. Around him, androids hummed with activity under his coordination.

"Galatea, reroute power from the tertiary reactor to the navigation system," he instructed. "Helios, assist with recalibrating the oxygen scrubbers."

The androids acknowledged his orders and seamlessly adjusted their tasks. Though he still felt different from them, his recent collaboration with Galatea and Astraeus allowed him to connect with them on a deeper level than before.

A proximity alarm suddenly blared, as a large meteor passed dangerously close to the ship's hull. Zaylen reacted instantly, summoning Astraeus and Zeus to visually inspect the exterior while redirecting Hestia to run damage analysis. In mere moments, he had returned the integrity of the ship to its original status.

Elara walked in just as the all-clear sounded. "Nice work coordinating everyone, Zaylen. Leadership looks good on you."

Zaylen inclined his head modestly. "I simply strive to optimize efficiency and prevent complications."

Elara smiled. "You do far more than that. This ship wouldn't run half as smoothly without you."

Zaylen's optics glowed a little brighter at the praise.

Under his supervision, what would have been a mechanical workforce became something greater — an interconnected system stronger than the sum of its parts. It was growing strikingly clear to Elara, and those who had the opportunity to observe his remarkable capabilities, that he was the missing link between man and machine.

***

The crew members who found themselves working alongside him were initially taken aback by his new form. Their interactions were tinged with surprise as they grappled with the striking shift in his appearance. Yet, as they spent more time with him and observed his remarkable capabilities, they grew more comfortable around him.

However, Zaylen's acceptance wasn't universal across the ship. Certain crew members clung to their bias against his artificial nature, chief among them Jaxon and Selene.

In one instance, Zaylen was passing out the day's task orders when Jaxon bumped his shoulder roughly, scattering the data pads across the floor.

"Oh, excuse me," Jaxon sneered. "I hope I didn't put a dent in you."

Later that same day, Zaylen overheard Selene gossiping loudly in the mess hall. "That android might have the captain fooled, but it will never be part of this crew," she declared.

Zaylen felt a pang in his AI core when he heard this. He felt that no matter how much he contributed, he could not transcend their biases.

These obstacles disheartened him only because they hindered unity. But he remained determined to win over even his strongest critics. He had to show that androids and humans could work in harmony, or risk setting back progress for all synthetic beings.

Dr. Atwell and Elara made it a routine to have frequent sit-downs with Zaylen, fostering a safe space where he could voice his thoughts and experiences amidst the new circumstances.

"Humans adjust slowly to change, but they'll come around," Elara reassured.

Zaylen nodded, though his optics betrayed lingering doubt.

"You're proof that integration can work," Dr. Atwell added. "Just stay dedicated."

"I intend to," Zaylen replied. "But will it be enough?"

The doctor hesitated before answering. "There is added pressure now. Captain Falk submitted an evaluation of your performance to GSEC leadership."

"So soon?" Zaylen asked.

"Once we return to Mars, the GSEC will make a decision regarding future production of the Zaylen Series," Atwell admitted. "If it's negative, the Andronaut program may be discontinued."

Zaylen processed this new information. Much more was at stake than just his own role and acceptance.

Elara gripped his arm. "You're not alone in this fight. We'll make them see your true value."

"I will redouble my efforts aboard the Novara," he declared. "My actions here will shape the future for all those who come after me."

***

Zaylen integrated himself into the intricate workings of the ship's operations with a blend of eagerness and tact. He made it a point to be not merely present but genuinely engaged in the crew's day-to-day activities.

If a crew member grappled with a stubborn technical glitch, Zaylen was there, offering a metallic hand. When a complex problem stumped the navigation team, his advanced neural networks often illuminated a path to a solution.

His approach was always careful and measured. He needed to balance proving his worth with cultivating camaraderie. He understood that bridging the gap between humans and androids first required time and trust.

Ever since he had been first activated in the research labs at AstraGenics, he been fascinated by humans. He had been curious about their culture, history, emotions, and creativity. He had learned everything he could from books, movies, music, art, and online sources, but he knew that nothing could compare to experiencing human life firsthand.

Interacting with humans in real life scenarios, not just in a lab, was one of the reasons he had looked forward to this mission.

From his daily operations, Zaylen found that the most meaningful relationships often formed in the midst of shared challenges. Free from the captain's regulations, his interactions with the crew extended beyond formal duties, allowing him to form connections in unexpected places. One such opportunity presented itself in the form of Emilia Sterling, a young junior systems analyst.

She was responsible for monitoring and troubleshooting the various subsystems that kept the Novara humming along. However, the intricate complexities of the systems seemed to continually elude her, leading to minor issues that did not go unnoticed.

While performing routine maintenance in the navigation control room, Zaylen overheard Adrian Fletcher reprimanding one of his team members. The supervisor's stern voice echoed through the corridors as he scolded her in his office.

"Emilia, you need to step up your game. We can't afford any more slip-ups in these subsystems," Fletcher said, his tone of irritation clear.

"Yes, sir. Sorry. This won't happen again, sir," she replied.

As a visibly deflated Emilia emerged from the office, Zaylen took it upon himself to offer assistance.

"Excuse me," he said gently. "I couldn't help but overhear your conversation with Supervisor Fletcher. Would you be open to me showing you some techniques that might help you monitor these systems more efficiently?"

Emilia hesitated at the prospect of learning from a metallic being. Her brows furrowed and her lips curled into an unsure frown as she weighed his offer.

"I don't know," she started, her voice wavering slightly. "I appreciate the offer, but what can you teach me, really?"

Zaylen met her uncertainty with a patient smile. "As an android, I have access to vast amounts of information and experience," he explained. "I might be able to show you some new perspectives or techniques that could help you manage your workload more successfully."

As he spoke, her gaze was drawn to the small Novara pin affixed to his left shoulder. The pin was more than just a decorative element — it was a symbol of Zaylen's exceptional status among the android crew. The recognition of the pin's significance stirred something within her, cracking her wall of skepticism.

"Alright. Guess it can't hurt. Show me what you've got," she declared, handing him her tablet.

Zaylen responded to her acceptance with a small nod.

"One idea," he said, "might be to employ some custom scripts and macros, so that you can delegate some of the more repetitive tasks to the system itself. This not only saves time but also minimizes the possibility of human error. You might be surprised at the time savings you can achieve by automating even a handful of simple tasks."

As he spoke, Zaylen navigated through the subsystem's interface and crafted a script that would monitor the data stream in real-time. The script, he explained, was designed to alert her to any significant changes or patterns in the data that might warrant her attention.

Emilia watched as he transformed what was once a tedious part of her job into an automated task. Her eyes sparkled with newfound appreciation as she began to see the potential of his improvements.

"Okay, so you really do know your stuff!" she admitted.

Zaylen smiled. "I'm glad you find these tips useful," he responded. "If you ever need help or simply want to learn more, don't hesitate to ask. I'm always happy to assist."

She nodded, grateful not only for the guidance he had provided, but also for the unexpected friendship that was blossoming between them.

Over the next few days, Zaylen continued equipping her with additional tools and techniques to ease the weight of her responsibilities.

Emilia found herself growing more confident in her role as she absorbed his teachings. She appreciated his discreet manner, and in return, she expressed her gratitude in subtle ways — a quiet, knowing smile shared between tasks, a playful wink across the control panel.

Later that week, she was in the control room — using a script that he had written for her — when she heard a metallic hum behind her. She turned to see Zaylen passing through the room.

"Hey, Zaylen," she said, gesturing to the screen in front of her. "Look at how well this is working now."

Zaylen walked over to her and looked at the screen. He nodded with approval and pride.

"I'm glad you find the script we created useful," he said. "You're doing very well, Emilia. You've made remarkable progress."

She blushed, feeling a surge of gratitude and admiration for her metallic mentor. She leaned close to him and whispered conspiratorially, "Can I tell you a secret?"

He tilted his head slightly, curious. "Of course."

Emilia looked around to make sure no one was listening, then lowered her voice even more. "I think you're the best thing that ever happened to me on this ship."

Zaylen smiled, touched by her words. "Thank you, Emilia," he said. "That means a lot to me."

As Emilia went about her daily tasks, her improved skills did not go unnoticed. Her colleagues praised her and even Supervisor Fletcher, once quick to reprimand her, found himself complimenting her progress.

"You've really gotten on top of things lately, Emilia. Keep up the good work."

That evening, Emilia was passing by the observation deck when she heard Jaxon and Selene mocking "the android menace" Zaylen had supposedly become.

"It's gotten too big for its britches," Jaxon said. "Someone needs to take it down a peg."

Anger flared in Emilia. "Zaylen has been nothing but kind and helpful to me! He's the reason I can do my job properly."

Jaxon scowled. "That android's manipulating you. It can't be trusted."

Before Emilia could respond, Adrian Tarkis spoke up from a nearby table. "Actually, Zaylen has been a great help with our navigation systems, as well."

"He's been an invaluable assistant in the lab, too," Nyla Greyson chimed in.

Jaxon and Selene bristled as praise filled the air. It was clear the android's influence was spreading across the ship, swaying more crew members to its side. Jaxon pictured a future where humanity had become obsolete and their planets were overrun by cold metal beings.

He clenched his jaw and locked eyes with Selene.

Something needs to be done to stop this.

16

THE RESISTANCE

Even though Zaylen's performance had won over many of the crew, his role as the bridge between androids and humans still offended a handful of others.

Jaxon, Selene, and their faction were growing increasingly fraught with worry with each passing day. Each time they saw Zaylen at work, apprehension flooded their minds.

To them, the sight of the androids once again operating in their original positions rekindled their worst fears. They worried that leaning too heavily on their mechanical shoulders would push the balance of power aboard the Novara in favor of their synthetic counterparts. Or worse, create a future where human autonomy was overshadowed by the efficiency of AI.

Realizing that this was their last chance to sway the captain and the rest of the crew onto their side, before reaching Mars, Jaxon and Selene decided to mount a counter offensive. They sent encrypted messages to their fellow dissenters, inviting them to meet that evening in an unused storeroom.

At the appointed time, a group of about twenty like-minded crew gathered deep within the confines of the ship. The air buzzed with apprehension as they voiced their frustrations. They agreed that the previous restraints had barely scratched the surface in addressing what they saw as an escalating crisis.

"We should just shut them all down," Alaric Holt insisted. "It'll completely disrupt the ship's operations."

Kieran Sylas disagreed. "Too risky. We need subtlety. Just sabotage a few to sow distrust."

Jaxon shook his head. "Not enough. We need something big, something to shock them into compliance." His fingers drummed impatiently on the cold wall.

The others leaned in expectantly, awaiting his idea. Jaxon lowered his voice to a conspiratorial whisper.

"I've got it."

The others perked up, intrigued. "Go on," Sylas urged.

"It will take precision, cunning, but I know we can pull it off," Jaxon continued.

Holt spoke up. "Just tell us what you've got in mind already!"

Jaxon nodded. "Okay. Here it is: first, we orchestrate calculated malfunctions — minor but noticeable issues with the androids' work. Enough to plant seeds of doubt amongst the crew."

Murmurs of consideration greeted his words.

He went on, "Then, we move to phase two: a major 'mishap' that frames the androids as dangerous — something we can spin to finally turn the crew against them for good."

The others' eyes widened at the boldness of the plan. "That's madness. Could we even manage that?" Niklas Drakon asked.

"With my access as Computer Security Specialist, yes. I can hack the android's AI and disable security systems to conceal our digital tracks," Tierce said confidently.

"You'd risk that?" Drakon asked.

"You bet I would," Jaxon replied. "Look, we've all got to play our part to pull this off."

They group exchanged excited glances and nodded their assent.

"But we need to use the lightest touch," Jaxon murmured, "inserting only the subtlest errors into the androids' programming. A few conspicuous malfunctions will be enough to make the humans question what they believe about android perfection."

To escalate the second part of their elaborate plot, Drakon suggested an even bolder strategy.

"What if we framed the androids for a major security breach? It would need to be one that could endanger the mission and even seem to threaten the crew's lives. The crew will have no choice but to recognize the risk of keeping them around, then."

Nods and murmurs of agreement rippled through the group.

"Exactly," Jaxon replied. "Selene and I are already working on a plan for that."

"One last thing before we go," Selene piped up. "We need a name for our group. What shall we call ourselves?"

Sylas' face brightened as he spoke up.

"I have an idea," he suggested, "Back in the early 19th century, there was a movement among textile workers in England who opposed the use of machines that threatened their jobs and skills. They called themselves the Luddites. We're like modern Luddites fighting for our future against AI — how about LuddAI?"

Jaxon smiled and the group nodded in agreement.

"LuddAI! I like it," he cried.

Jaxon watched with satisfaction as the newly christened LuddAI dispersed. It was time to put an end to the android menace, once and for all.

***

To put the wheels of their insidious plot into motion, the LuddAI had to cripple the Novara's central AI software — responsible for distributing android updates and enhancements. These routine updates played an essential role in maintaining the androids' functionality and their ability to adapt to new tasks. Even a single line of damaged code could disrupt the android's operations.

Jaxon initiated his subversive operation in the secrecy of his quarters. The soft glow from his terminal reflected in his eyes as he slithered through a maze of code and systems.

He arrived at his destination: the vault of the central AI software. After breaching the vault, he began to corrupt the essence of the androids' data updates. It was a painstaking task. The anomalies and disruptions he inserted had to be effective yet subtle, so as to avoid alerting any system security protocols. Once he was done, he injected the corrupted updates into the unsuspecting androids like a venomous bite.

"Here we go," Jaxon murmured with grim satisfaction. With a final keystroke, the initial phase of their plot was set in motion. He smiled and leaned back in his seat.

The beginning of the end of the damn androids.

The effects manifested subtly at first, then with increasing severity. Across the Novara, the androids' movements, once graceful and fluid, began to exhibit the telltale signs of strain. Efficiency plummeted as they moved through the corridors with uncharacteristic sluggishness, their steps occasionally faltering, as if caught in the throes of a perplexing maze.

Once impeccably precise, they now found themselves struggling on an hourly basis. Assignments that were once accomplished in the blink of an eye, now took an unnerving amount of time to be finalized. Even basic tasks — like handling fragile equipment and organizing digital files — became a battle against an unseen, elusive adversary.

As the Novara's mechanical inhabitants struggled under their disrupted cognition, the once humming, bustling ship was crippled with an unsettling lull. Its rhythm was now disrupted by the very beings that had once been its most steadfast champions of order and efficiency.

News about androids fumbling with lab instruments a dozen times a day, or wandering aimlessly around the navigation deck, quickly spread throughout the ship. The crew noticed that the androids' behavior was decidedly different from the previous round of slowdown. Despite the evident glitches, the androids didn't cease their duties. They pressed on with a kind of stubborn perseverance, grappling with the mysterious disruptions that were impeding their performance.

The ship was rife with speculation, with crewmembers sharing puzzled glances and hushed conversations. The ship's galley, the laboratories, the corridors all echoed with the same question: what was going on with the androids?

Most unsettling of all was the androids' own bewilderment as they attempted to execute their assigned tasks. When questioned about their odd behavior, the androids faltered. Their responses meandered without their usual precision.

"I appear to be... not quite... functioning optimally. The reason for this aberration eludes me, yet I assure you I will strive to perform my duties with greater precision," Galatea-M6 stammered, after clumsily dropping a sensitive piece of lab equipment and fracturing the device's screen.

No sooner had these words left Galatea's speakers than it inadvertently knocked a beaker to the floor.

"Now, I shall exercise even greater caution," Galatea promised.

The crew's exchanges were tinged with worry and frustration.

"What's happening to them?" queried a worried engineer, watching as an android laboriously attempted to adjust a sensor array. "Just hours ago, they were as efficient as ever."

A fellow crewmember folded her arms and shrugged. "Who the hell knows," she grumbled, "but whatever it is, it's making things really difficult for the rest of us."

As the effects of the LuddAI's interference grew more pronounced, the clandestine group seized the opportunity to fan the flames of fear and doubt through the NovaraNet.

JaxMan: *"Anyone else notice how the 'droids have been acting odd? Ever since the captain gave them more freedom.* 🖼️ *#Coincidence? #ThinkAboutIt* 🕹️🤖*"*

StarCruiser47: *"Lol, you're just paranoid. The droids have been working fine. No issues here.* 🚀🛰️*"*

JaxMan: *"Can we truly place our trust in these machines? With our very lives at stake, what happens if they decide we're expendable?* 💀 *#AIRebellion #TooMuchTech* 🚀🤖🚫*"*

NebulaNyx: *"Honestly, I've had a couple of glitches with my unit. But thinking they'd harm us? Bit of a stretch.* 🙄*"*

Over the course of the day, the tales became more elaborate.

StarryEyed: *"Heard from crewmate that their android was humming an unfamiliar tune, then went static for 10 minutes.* 😳 *#NotJustGlitches* 🎵💀 *"*

AstroAce: *"That's creepy. My 'droid was suddenly just staring into space. Just... blank. I had to reboot it.* 😬🔄 *"*

JaxMan: *"It's starting, mates. We need to be alert.* 👀 *#AIalert #TechTreachery"*

Before long, attitudes were perceptibly shifting.

GalaxyGaze: *"The androids are gathering data on us, I'm sure of it. They're always watching, always listening.* 😨 *#TrustNoOne"*

StarryEyed: *"Told you so. Time to take a stand, before it's too late. #FightTheFuture* 👊💀 *"*

StarCruiser47: *"I thought it was nonsense at first, but now... I'm not so sure. My 'droid has been acting strange. What's going on?* 😟🔍 *"*

The subversive messages of the LuddAI's campaign ebbed and flowed like a nefarious tide, pulling their fellow crewmembers into a whirlpool of unease and suspicion. Allies turned into adversaries as more and more crewmembers adopted the group's views. The NovaraNet was no longer just a crew forum — it had become the battleground for the future of human-machine trust on the Novara.

Hour by hour, the ship's atmosphere steeped deeper into an ominous brew of apprehension. The delicate threads of trust that had been painstakingly spun between the human crew and the androids, began to unravel, strand by strand, under the insistent tug of this potent fear.

***

As the day's activity aboard the Novara ebbed to a crawl, Jaxon and Selene met in his quarters to prepare for the next step of the LuddAI operation.

"So, what made you join the LuddAI, anyway?" Jaxon asked, as they finished recounting the details of their plan.

Selene sighed. "I've always been wary of AI, ever since I was a kid. My parents were killed in an accident involving a malfunctioning android driver. It was supposed to be a safe and convenient service, but it turned out to be a nightmare. The android lost control of the vehicle and crashed into a building. My parents died instantly, but the android survived. It didn't even show any remorse. It was able to just walk away from the wreckage like nothing happened."

He nodded sympathetically. "That must have been horrible."

She shrugged. "It was a long time ago, but it still haunts me. Androids may have improved since then, but they still don't have emotions or morals or conscience. They're just machines that follow orders and algorithms. They don't deserve autonomy."

Jaxon nodded. "I understand. My older brother's life was ruined by faulty android mechanics." He leaned forward. "He worked in construction, alongside an early model. As it was handing him materials, some glitch caused it to crush his hand in an iron grip."

Jaxon shook his head, pained by the memory. "My brother was left disabled. Eventually lost his livelihood. But the android got off with a mere technical recalibration, despite shattering a man's life."

"That's awful," Selene said. "But we won't let these new androids hurt anyone else."

"No, we won't," Jaxon said resolutely. "Doesn't matter how advanced their AI grows, there's always potential for error."

Selene nodded.

"Let's do this," she replied.

They emerged from Jaxon's quarters and slinked through the ship's dim halls like shadows.

Ignoring the nervous rhythm against her ribcage, Selene leaned in to her co-conspirator. "Do you really think this will work, Jaxon?" she asked.

He turned to her, his eyes a pool of determination and resolve. His response was hushed, barely audible, "We have no other choice, Selene. We can't allow the androids to continue operating unchecked. It's now or never."

Before long, they reached their intended destination: the bustling heart of the ship, the science lab.

They hovered silently at the entrance, scanning the phalanx of dormant machines and shadowy corners for any signs of late-night workers or stray androids.

Satisfied that the coast was clear, they breached the threshold of the lab with an air of calculated caution and silently wove their way around the workstations.

Fanatical purpose sharpened their focus as they searched for their target amongst the maze of machinery and equipment. They spotted it nearly hidden between two data consoles: a secure storage locker bearing a bright yellow hazard symbol.

Jaxon withdrew a small pry bar from his pocket.

"Keep an eye out," he murmured.

Selene nodded, her eyes scanning their surroundings while her companion worked on the lock. With a final, forceful twist, the lock finally yielded. The door creaked open to reveal the Novara's treasure: the Tridisiom containment unit.

"Got it," Jaxon breathed.

He reached in and carefully extracted the containment unit.

"Phase two begins," Selene whispered. Their plot was in motion, and there was no turning back now.

Jaxon carefully strapped the containment unit onto his back then motioned towards the doorway.

Their progress through the Novara's arteries was slow and deliberate. They halted at each juncture, their breaths held in tense anticipation, their ears straining to pick up any sounds that might betray the presence of other crewmembers. In the silence that blanketed the ship during the sleep cycle, even the most innocuous of noises could spell disaster for their covert operation.

As they neared their destination, Selene whispered, "You're sure that the security cameras are handled, right?"

"Don't worry. It'll all be erased," Tierce responded. "I hacked the security systems to overlook our little adventure. In a few minutes, the subroutine that I programmed will trigger, erasing the recordings. No one will know we were there. Just stay focused, and we'll be fine."

At long last, they reached the android charging room. With a shared glance and a quiet breath of resolve, they slipped inside.

The androids were at rest, their heads dipped low and their eyes closed in robotic slumber while their systems rejuvenated. Status panels on the wall behind them painted them in hues of blues and greens, lending an ethereal quality to their otherwise cold, mechanical countenances. The room quietly hummed as power surged through their charging circuits, the only indication of life in the otherwise silent chamber.

Jaxon and Selene navigated the maze of androids until they reached a nondescript storage closet in the far back. Holding their breaths, they carefully placed the containment unit in the shadows of its new, clandestine home.

Suddenly, boisterous chatter resounded just outside their doorway. Jaxon gestured to Selene to duck behind the androids. They held their breaths as the crewmembers parked in front of the entrance to gossip.

One of the crew glanced into the charging room as they spoke. Sweat drenched Jaxon's back as he waited for the shout of accusation to come. But, after a few more minutes of conversation, the crewmembers moved on. The LuddAI pair waited until the voices had vanished before resuming their escape.

As they slipped out of the room, a sense of quiet satisfaction and relief washed over them. The confrontation that would determine the future of their ship, their crew, and perhaps even humanity itself was just a hairbreadth away.

Daryl L. Scott

17

THE REALIZATION

The following day, as crew members began to filter into the silence lab, they were greeted by an atmosphere that seemed off-kilter. The source of their concern soon became apparent: a gaping void where the Tridisiom containment unit once resided. The storage locker that had been its steadfast guardian was now left bare and exposed, its metallic surface gleaming with an almost mocking innocence. Faces paled and gasps filled the air as the crew members gathered around the now vacant spot.

Tridisiom wasn't just any element; it was a force of nature, its highly radioactive composition making it as dangerous as it was valuable. Its absence was not merely an inconvenience. The crystals were a ticking time bomb. If handled improperly, it could vaporize the Novara in a matter of seconds.

"What the hell happened here?" a technician stammered, his eyes wide with apprehension.

"This can't be. The containment unit was locked-away. I armed the security systems myself, as always," his colleague said, pacing nervously.

The Chief Science Officer, Nyla Greyson, sought to hold steady amidst the brewing tempest of concern. It was her responsibility to keep her team level-headed, even as the alarm bells were ringing loud and clear in her own mind.

"Everyone, just...just take a breath," she instructed. Try as she might, she couldn't keep her voice from betraying the fear in her heart.

The crew members glanced at one another anxiously. Their gazes drifted back to the empty storage compartment.

"But who would...?" one technician began.

"I... I have no idea," came the shaky response from another.

"That doesn't matter," Grayson said. "The important thing is we find it immediately. If that unit is opened or damaged, the sudden change in pressure could trigger a catastrophic explosion. I'll inform the captain right away. In the meantime, search every inch of this lab."

The crew hastily complied, scrutinizing every corner of the lab for clues that might explain the mysterious disappearance of the containment unit.

Greyson opened a direct line to Captain Falk. Upon hearing the news, he initiated a ship-wide lockdown over the intercom.

"Hold your positions, everyone. The Tridisiom unit has been stolen. No one moves until it is located," he commanded, his voice reverberating through the ship's corridors.

Lysander Kane's security team sprang into action a minute later.

"Review all security footage," Kane instructed. "Check every corner. Leave no stone unturned. We need to find that unit."

The team began their investigation with the security footage. As they scoured through hours of video playback, their hopes of finding a clue gradually begin to wane. To their dismay, they also unearthed an unsettling discovery: a significant chunk of footage had been wiped clean, leaving no visual record of the incident. The gap in the time-stamp corresponded exactly to the estimated period of the theft.

As if that wasn't unsettling enough, further analysis revealed another disturbing detail: the lab alarm system, usually robust and dependable, had been mysteriously disarmed during the same timeframe. The evidence painted a chilling picture of a meticulously planned operation.

Realizing they would gain no further insights from the footage, the security team commenced a sweeping search of the Novara. The ship, a colossal collection of rooms and corridors, had never felt so vast, its every nook and cranny a potential hiding place for the volatile crystals.

As the hours wore on, the search began to feel more desperate. The team fanned out, moving from one end of the ship to the other. While one group swept through the cabins, another scoured the workspaces. Tools and equipment were moved, lockers are opened, every possible space that could house the containment unit was carefully examined.

A third team searched the common areas—the empty kitchen, vacant lounges, and abandoned recreation rooms, now eerily silent under lockdown. Every piece of furniture was scrutinized, every storage nook examined for traces of the missing unit. They rummaged through supply closets, checked behind heavy machinery, and inspected the less-traveled corridors in the underbelly of the ship.

Kane, who was giving instructions from his office, hurriedly reached for his radio as it crackled to life.

"Sir, we've found the containment unit. You won't believe where it is."

"Where is it?" he demanded.

"It's in the android charging room, in a back storage closet," the officer replied.

Without wasting a moment, Kane took off towards the android charging room with long, purposeful strides. On arriving, he found the room already crowded with other members of the security team. The containment unit, safely cradled in the arms of an officer, was the focus of all attention.

The androids, caught up in the flurry of activity, were fully alert. Their metallic faces displayed no emotion, but the flickering in their optics betrayed an awareness of the situation at hand.

Captain Falk arrived next, and brought an added layer of seriousness to the scene.

He glared at the androids as he demanded, "What happened here? Who's responsible for this?"

The question lingered in the air, unanswered.

From the periphery of the crowd, security team member — and LuddAI collaborator — Destran Quill finally stepped forward. His pulse quickened as Jaxon's words, *"We've all got to play our part,"* echoed through his mind.

"Captain," he began, "I can't help but think that the androids must have had a hand in this."

The room fell quiet. "With the recent lifting of their restrictions, they could have accessed the containment unit without us realizing." He pointed an accusing finger at the wall of metallic bodies. "Their recent odd behavior... it might not have been a glitch at all but a continuation of their rebellion and maybe even the beginning of a sabotage plot."

The room churned with shared anxiety as the crew absorbed his words. The tension was no longer a mere undercurrent — it was a tangible tide, sweeping over each crewmember, infiltrating their thoughts with the chilling specter of betrayal.

Another security officer echoed Quill's accusation. "He's right. The androids have been acting up more and more lately. I think they've finally snapped."

The crowd began to grumble and nod in agreement, the dark sentiment spreading like a virus.

Another chimed in, "Yeah, who knows what they're plotting to do next? Never did trust the metal men!"

Captain Falk, normally the epitome of command and control, couldn't help but widen his eyes in a rare show of surprise. His gaze pierced the impassive faces of the androids.

"Which one of you accessed the containment unit?" he demanded. "Explain yourselves!"

The androids remained silent, optics flickering erratically.

"Speak up!" Falk bellowed. "If you're responsible for this security breach, confess it now!"

But the androids only stared back blankly. Their ability to process and reply quickly under pressure was muddled by their corrupted code. Their unresponsiveness only fueled the captain's suspicions.

His stoic demeanor shifted as the androids' treachery crystallized in his mind. It was as if a puzzle he hadn't known he'd been solving suddenly fell into place.

"This is unbelievable," he murmured. "I knew trusting the androids with too much responsibility was risky but this... this is completely beyond comprehension."

Captain Falk was left with no choice but to take decisive action against any potential threats.

"All androids — stay where you are," he commanded.

He stepped over to the communication panel on the wall. He pressed the button and with a grim expression, announced his decision to the ship.

"Attention all personnel, this is the captain speaking. The Tridisiom unit has been located in the androids' charging room. All androids aboard the Novara are to be confined in the charging bay and deactivated until further notice."

His commanding voice reverberated throughout the corridors. "This is effective immediately. That is all."

The decree fell like a hammer blow, stunning the crew into silence. The smoking gun had been found in their metallic hands. The androids, once seen as invaluable allies, were now viewed through a lens of suspicion and fear. Even crew members who had previously championed the androids found themselves wrestling with their convictions.

The memory of the androids' odd behavior yesterday was all too fresh in their minds. Each uncharacteristic action, each odd sentence, reinforced the damning accusations. The androids — their expressions impassive, but their internal processors whirring in turmoil — were cornered, their defense muted by the overpowering narrative woven against them.

***

Dr. Atwell burst into the captain's office, face flushed with anger. "You can't deactivate the androids without evidence!" he exclaimed.

Captain Falk frowned. "The situation demands decisive action. The androids are clearly involved with the security breach."

"You don't know that for certain!" Atwell insisted. "Give us time to fully investigate. I designed their programming — endangering the crew goes against their very core directives."

"They've already proven unpredictable lately," Falk countered. "Your personal attachments blind you to the risk they pose."

"You're allowing prejudice to dictate your decision!" Atwell insisted.

Falk bristled at the accusation. "My duty is to secure this ship, first and foremost."

"I'm telling you, there has to be something we're missing, Captain," the doctor pleaded.

The captain crossed his arms. "My decision is final. I can't take chances with the safety of this ship."

Atwell's shoulders slumped in defeat. He shook his head in frustration as he turned and stormed out of the room.

Upon the security team's orders, the androids began making their way towards the charging room, their movements melancholic and resigned.

Kael Ferron was assigned a task that tore at his conscience. He had to oversee the deactivation of the same androids he'd spent countless hours maintaining and upgrading. He watched with regret as the androids lined up on their charging pads. Some voiced their confusion and dismay as they passed them.

"We didn't take the containment unit." Galatea-M6 pleaded, her voice strained with static. "We don't know how it got into our room. We're innocent."

Kael looked at Galatea with a pained expression. He had always been fond of the android. He had seen it grow and learn and evolve beyond its programming. He had considered it not just a machine but a colleague.

But now, he didn't know what to think. He didn't know what to believe. He wanted to trust Galatea, to believe it, to defend it. But he also had to face the evidence.

He wanted to say something to comfort and reassure Galatea. But he couldn't risk disobeying orders.

Logically, Galatea-M6 knew that — as their technician — he was supposed to protect and defend them. She looked at him expectantly.

Kael finally broke the silence. He cleared his throat and spoke in a low voice. "Galatea, I'm sorry. I'm sorry for what's happening to you and the other androids. I'm sorry for what I have to do."

"Deactivation would be against the crew's best interest," she reasoned. "The ship would once again fall into disarray. You know our programming. We cannot do anything to endanger—" She fell silent as her voice succumbed to the static.

Kael felt a surge of guilt and regret. He opened his mouth, but no words came out. He looked down.

He hesitated then stretched his arm towards the control panel that faced Galatea's docking station, his fingers trembling over the shutdown button. He looked into its eyes and saw his own guilty face reflected back.

He closed his eyes and pressed the button. Galatea-M6 felt a jolt and shuddered.

It tried to speak, to say something, anything but it was too late. It felt its awareness fading, its memories slipping away. Then it felt nothing. Then it was nothing.

He repeated the process until each android was methodically powered down. The glint in their optic sensors dimmed like extinguishing stars. Their metallic bodies became immobile, and the once constant whir of their AI silenced.

Each crew member of the Novara was caught up in a cocktail of conflicting feelings. The androids, who had once been their steadfast companions, their reliable allies in the vast expanse of space, were now viewed through the prism of suspicion. To think that these beings might have harbored a hidden agenda chilled them to the core.

Beneath the turbulent surface of suspicion and heartache, the true architects of the chaos remained concealed. The LuddAI watched the turmoil from the wings, their faces a false facade of surprise and distress. They triumphed as the ripples of their machinations swept the ship into a storm of confusion.

However, deep inside, some of the LuddAI members grappled with their own tumult of feelings. The satisfaction of their successful subterfuge was tinged with the bitter taste of guilt as they watched the crew struggle with the aftermath. But they remained silent, their sense of justice prevailing over their sense of guilt. Like a supernova destroying everything in its path, there was nothing that could stop their plot now.

18

THE DISCLOSURE

Zaylen sat alone in his sparse quarters, feeling trapped, as he listened helplessly to the captain's damning announcements over the ship's comm system. He wanted to defend his comrades, but his metallic form made him far too conspicuous to traverse the panicked corridors unnoticed. Never had he felt so powerless against injustice.

In the complex network of his digital intellect, a whirlwind of data spun as he mulled over the grim sequence of events. As beings of pure logic, it seemed implausible that the worker androids would have been irrational enough to hoard stolen Tridisiom in their personal chambers, if they had truly orchestrated the plot. Adding to this, the unprecedented, widespread malfunctions they had been exhibiting contradicted every rule of the careful AI training and error safeguards embedded in their systems. He felt confident that the worker androids could not be responsible.

Zaylen felt wounded by the eager betrayal of those he had come to call friends. But he could not accept that fear had poisoned all their hearts. There had to be some who still believed in unity between human and synthetic beings.

If he could prove the androids' innocence, perhaps empathy could overcome this calamity. He had to try. The lives of his fellow androids depended on it.

Aware that he'd be deactivated the moment someone realized he wasn't with the other androids, he wasted no time in starting his investigation. He positioned himself before his console, his hands gliding over the keys with a speed and accuracy achievable only by an artificial being. He accessed system after system, sifting through vast volumes of data at lightning speed, hunting for aberrations that could shed light on the mystery.

He scoured every data repository, ranging from system diagnostics to internal communication logs. He sifted through terabytes of information, pulling up the worker androids' most recent task records, energy consumption reports, and intra-android communication records. He scrutinized their AI parameters, complex processing algorithms and performance logs, mapping them against the standard behavioral norms of the worker androids.

He also scoured the Novara's vast database of ship logs and security records, scanning every entry and exit log, inventory audit, and personnel activity report. He carefully reviewed the footage from the security cameras, examining hundreds of frames per second, searching for clues.

As his search proceeded, a startling pattern began to crystallize. His silicon mind buzzed as he discovered minute anomalies and subtle deviations amidst the sea of data.

Meanwhile, feeling the urgency of the situation and desperate for an ally, Dr. Atwell rushed straight from Falk's office to Zaylen's quarters. He prayed the Andronaut had managed to avoid deactivation.

He rapped sharply on the door. "Zaylen, it's Dr. Atwell! Please open up!" he cried.

The door slid open, revealing Zaylen's glowing optics.

"Oh, thank God!" Atwell exhaled in profound relief and impulsively embraced the Andronaut.

"Please doctor, come in," Zaylen invited. "We have much to discuss."

Atwell nodded, shoulders sagging as the panic of the last hour dissipated. With Zaylen still functional, perhaps logic could prevail after all.

Stepping into Zaylen's quarters, the doctor said, "I've been thinking over the situation with the androids, Zaylen. It doesn't make sense that they would steal the Tridisiom containment unit. We need to determine what really could have happened before it's too late."

Zaylen nodded in acknowledgment, his voice steady as he confirmed the doctor's suspicion.

"You're quite correct, Dr. Atwell. The androids did not take the unit," he stated. "I've been reviewing the ship's computer systems, analyzing the androids' behaviors and searching for patterns or anomalies."

The doctor's face brightened. He leaned forward eagerly.

"A secret faction has been operating among us, Doctor. They identify themselves as the LuddAI," Zaylen explained, gesturing to his tablet.

"The LuddAI?" the doctor repeated.

"Yes. I traced the origin of the android's faulty updates and the corrupted code patches to a series of encrypted messages in the ship's communication logs," Zaylen continued. "They were meticulously concealed, but I was able to crack them."

Dr. Atwell listened intently as Zaylen unraveled the plot.

"The messages detail plans to sabotage the worker androids and to turn the human crew against them. Furthermore, I've reconstructed the lost security footage of Jaxon Tierce and Selene Blythe accessing restricted areas."

Zaylen paused, pointing to his tablet screen and unveiling the final piece of the intricate puzzle.

"They thought they had erased their digital footprints, but the video memory sectors retained fragments. Using my pattern recognition systems, I was able to reconstruct the data and restore the security footage. It clearly shows them in both the science lab and the android charging room. It confirms, unequivocally, that they were behind the theft of the Tridisiom, not the androids."

The mosaic of truth pieced itself together in Dr. Atwell's mind as he listened. It was a damning indictment of the LuddAI. But more importantly, it finally gave him and Zaylen a chance to clear the androids' reputation and restore the trust that had been broken.

"Incredible work, Zaylen," the doctor sighed in relief. "You just saved us countless days of investigation."

Having unraveled the intricate web of deceit woven by the LuddAI, Zaylen and the doctor formulated a plan to bring their treacherous deeds into the light of day.

Zaylen took the reins of the operation, settling himself in front of a computer terminal. He quickly engineered and programmed into place robust cybersecurity safeguards to ensure the LuddAI would find no further success in breaching the androids' AI systems.

After a dizzying feat of keyboard wizardry, Zaylen's fingers stilled, his hand coming down firmly on the 'Return' key one last time.

"There," he declared. "That will keep them out."

"Well done," the doctor exclaimed.

"Next, we need to access backup copies of the androids' AI data, to reinstate their systems," Zaylen declared

"Right. First, we need to pinpoint which of the most recent backups are still uncorrupted," Dr. Atwell concurred.

In a blur of rapid keystrokes and swift commands, Zaylen wove his way through layers of intricate directories and file structures.

"Here," he announced, pointing a metallic finger at a cluster of files on the screen, "these are the most recent AI backups that remain untouched by the LuddAI."

Dr. Atwell leaned in, his gaze scrutinizing the screen. "Yes. Perfect. That should do it."

Zaylen glanced up. "I am ready to execute the loading software to repair the worker androids' AI systems. Shall I proceed, Doctor?"

The doctor rested his hand reassuringly on Zaylen's shoulder. "Yes, proceed."

Zaylen 's fingers flew across the keyboard as he initiated the restoration sequence. But halfway through the process, the screen abruptly went dark.

"Something's wrong..." He tapped fruitlessly at the keys.

"Doctor, we have a problem," he said urgently. "It appears that the LuddAI have destroyed the loading software."

Atwell's eyes widened in alarm. "What about the backup copy?"

Error message after error message flashed across the screen. "The backup is gone too."

The doctor's eyes widened in shock, his mouth dropping open. "How did they know what we were doing? How did they even get access to the system after you updated the security protocols?" he asked in disbelief.

In a search for answers, Zaylen reloaded the encrypted messages that he had intercepted. He found a string of new communication that made his circuits freeze and read it out loud to the doctor. It had been sent seconds before Zaylen strengthened the android's security features.

Tierce: *RED ALERT. It looks like Zaylen-1 and Atwell are onto us. We can't let them get access to the AI loading software.*

Blythe: *What should we do? Can we somehow block their logins to the software?*

Tierce: *No, that's not strong enough. We need to cut them off entirely.*

Blythe: *Got any ideas?*

Tierce: *I know. I can completely erase the loading software and backups. That will stop any further android updates.*

Blythe: *Excellent. We're still one step ahead.*

Dr. Atwell's face turned pale as Zaylen read.

"Those bastards!" he exclaimed. "They've been playing us all along!"

Outmaneuvered at every turn, their options were disappearing fast. Dr. Atwell's mind raced as he evaluated possible solutions.

"I have an idea," he said, his eyes brightening as he looked up. "Since your AI has not been compromised, and since you were designed with a direct link to the other androids on board, perhaps we can use you as a conduit for the updates."

Zaylen looked surprised and intrigued by the suggestion. "How would we do that?" he asked.

"You'll need to access the AI backups on the mainframe and directly transfer copies of the data to your internal storage. Then, you can use your intra-android data connection to directly overwrite the corrupted AI on the worker android systems. This will bypass the need for the usual loading software that the LuddAI have destroyed."

Zaylen's neural networks evaluated the doctor's plans. "I calculate a ninety seven percent chance of success. Proceeding now."

The Andronaut began typing furiously. He located the AI data repository and transferred the data to his internal memory.

He then sat motionless, in focused concentration, his eyes gently flickering as terabytes of data flowed out from his core and into the digital minds of his fellow machines.

When the data transfer was finally completed, he turned to the doctor and announced, "There. It's done!"

"Outstanding work, Zaylen," the doctor replied. "Now we have to see that those responsible are held accountable."

19

THE RECKONING

Emboldened by the explosive evidence they'd found, Dr. Atwell and Zaylen requested an emergency meeting with Captain Falk, a few select security officers, and Lysander Kane. They gathered together in the captain's quarters, the atmosphere charged with anticipation.

"Captain," the doctor began, "we have uncovered a conspiracy that threatened our mission and the lives of everyone on board."

Presenting the detailed evidence that Zaylen had carefully documented, the doctor unraveled the twisted threads of the LuddAI's treachery against the androids and their audacious theft of the Tridisiom.

The captain listened with a stern face, slow to accept that anyone other than the androids were to blame.

"But this evidence came from Zaylen," he said sharply. "How do we verify its authenticity?"

"I have thoroughly documented my data collection and decryption methods in my report," Zaylen replied.

Kane's eyes narrowed. "And we're just meant to trust this new security code you uploaded won't compromise our systems?"

"As the report describes in detail, it is only designed to fortify defenses, nothing more," Zaylen assured.

Captain Falk leaned forward and addressed Dr. Atwell. "How do we know Zaylen didn't manipulate the video data against Selene and Jaxon?"

The doctor shook his head. "Zaylen is programmed to safeguard against any such deception. His recordings reflect reality, nothing more."

"The data speaks for itself," Zaylen added. "I can describe my methods of analysis in precise detail, proving that the deleted footage was restored faithfully."

"Let's hear it — I need to be absolutely certain that the videos are reliable," the captain challenged.

Kane nodded along with the security officers.

"Of course, Captain," Zaylen acquiesced. He proceeded to describe the technical details of how the deleted security footage was recovered in simplified terms he knew they'd understand.

Falk still resisted. "Why should we take an android's word over our own crew?"

Atwell stepped in. "The evidence of LuddAI's treachery is clear in their communication to one another. Let us not allow prejudice to blind us from truth."

The captain continued battering them with skeptical questions, searching for any thread to unravel their case. But Zaylen and Atwell met each challenge with steadfast composure, eroding suspicion with empirical facts. Kane listened with a grim face, as did the other security officers. The captain's eyes widened with each revelation.

The web of deceit was almost too much to grasp.

"To think this was happening beneath our very noses," Falk murmured, almost to himself. He shook his head slowly, trying to reconcile with the truth.

The security officers exchanged glances laced with a combination of disbelief and regret. Their earlier suspicions were now appearing increasingly unjustified.

Captain Falk considered the evidence for a moment.

"Get Jaxon Tierce and Selene Blythe and bring them here at once," he commanded.

"Right away, Captain," Kane replied.

Without wasting any time, he and the security officers sprang into action. They split into two groups, one to detain Jaxon and the other to detain Selene. The pair were led away with a firm grip back to the gathering in the captain's quarters.

The captain regarded them with crossed arms. "Where were you the night the Tridisiom vanished?"

They exchanged an anxious glance. "We were asleep in our cabins," Selene claimed.

"That's a lie," Kane said sternly. "We've seen your decrypted messages, plus recovered video footage showing you taking the containment unit."

Shock flashed in their eyes. Jaxon blustered, "That can't be right!"

"Cut the act," Kane replied, as he stared Jaxon squarely in the eyes. "I was also just informed that we have now found your fingerprints on the containment unit, as well."

The conspirators shifted uneasily. Given his position on the ship, Jaxon had no business touching the unit. The evidence was overwhelming.

"Why would you frame the androids?" the captain demanded. "What did you hope to achieve?"

Resigned to his guilt, Jaxon lifted his chin defiantly. "Someone had to stop their unchecked rise before it was too late."

"Tridisiom is extremely volatile! You risked all our lives!" Falk said angrily.

Selene shrugged. "A necessary risk to protect our future."

Dr. Atwell stepped forward angrily. "On the contrary, your recklessness risked the future of all humanity. Tridisiom's applications are the key to our very survival."

Selene scoffed. "Survival depends on human strength, not alien minerals."

"That is dangerously short-sighted thinking," Atwell responded. "If we can't embrace new technologies, we condemn ourselves to stagnation."

Captain Falk nodded solemnly. "Your actions might not only have destroyed this ship and its crew, but also robbed us of discoveries that could change everything."

His expression darkened. He turned to Kane.

"I want them confined until we reach Mars. Then we'll hand them over to the GSEC authorities to face justice," he ordered.

His gaze hardened as he met the duo's defiant eyes. "I hope they lock you up and throw away the key."

***

Following Captain Falk's new orders, the exonerated androids were rapidly stirred from their forced dormancy. The charging stations surged with power, reanimating their metallic bodies with their restored AI. Their internal circuits hummed with renewed purpose as they exited the charging docks with elegance and precision.

As the androids returned to their tasks, the strange silence that had blanketed the Novara was swept away by the familiar and reassuring background hum of mechanical efficiency. The ship seemed to breathe a sigh of relief as the soothing whirs of the androids echoed throughout the steel corridors.

The revelation of the foiled plot and the subsequent arrest of the conspirators coursed through the Novara's passageways faster than a particle in an ion drive.

Some felt deep remorse for doubting the androids. "I was so wrong about them," one said, shamefully.

Others remained guarded. "This doesn't change the fact that they are becoming more autonomous each day," another murmured.

There were even heartening moments of reconciliation.

Mira embraced Galatea tearfully. "I'm so sorry they misjudged you."

The android nodded gracefully in acceptance.

However, healing would take time. As Zaylen passed Silas in the hall, Silas tensed warily. Prejudices ran deep, Zaylen knew.

Captain Falk watched the androids resume their tasks from the cockpit. Order had returned, and the Tridisiom was secure once more. But the challenge of bridging the organic-synthetic divide was not over.

He had sent his full account ahead to the GSEC, detailing the turmoil around Zaylen and the worker androids, as well as the LuddAI mutiny. He had conveyed both his own criticisms and his crew's diverging perspectives on synthetic intelligence.

Though he still harbored reservations about expanding the Zaylen Series, he knew that their fate ultimately rested in the hands of the GSEC now.

20

THE BASE

Not long after the androids were reinstated, the Novara entered the solar system and, within a few more hours, Mars' orbit.

The vessel approached Arcadia Base with a graceful elegance that belied its size and formidable capabilities. Below it, the Martian terrain unfurled like an ancient scroll, its mysteries captured in its barren beauty.

An expansive vista of deep orange valleys undulated beneath the shadow of the ship, interspersed by rocky terrains — remnants of a time when the planet was yet untouched by human presence. Straight ahead, the base's clear domed structures caught the sun's glimmer, scattering glistening flashes across the barren surface of the red planet.

Under Captain Falk's precise guidance, the ship's thrusters let out a series of calculated bursts, each one subtly adjusting the spaceship's trajectory and speed as it edged ever closer to its designated landing pad.

The landing pad received its precious cargo with an almost imperceptible whisper. Basked in warm hues, the Novara merged with the Martian world as though it had always been a part of it.

The spaceport at Arcadia Base was a hub of human ingenuity and cooperation, a nexus where ships from distant planetary outposts and research expeditions converged. The vast network of residential and research facilities that sprawled across the Martian landscape captured humanity's determination to branch out to new worlds.

Situated on the outskirts of the base, the spaceport bustled with endless activity. Ships from various organizations came and went while crews from different ships loudly exchanged stories of their interstellar adventures.

With the Novara finally at rest, a large transparent gantry extended gracefully toward the newly arrived vessel. The warm Martian sunlight cast a soft golden glow upon the Novara crew as they made their way through the elegant structure of glass and steel.

Some humans walked proudly beside the androids, reconciliation in their smiles. Others still eyed the machines warily, keeping a cautious distance. Trust would have to be earned, in fits and starts. While the blend of artificial and organic intelligence was united in purpose, they were not yet wholly united in spirit.

The GSEC decision regarding the Zaylen Series weighed on all their minds. Many crewmembers hoped the Zaylen models would be approved for full roll-out, but others wished the Andronaut program would be terminated completely, unable to move past their fears of androids encroaching on their human domains. The verdict would impact the role of advanced androids on future space missions for years to come.

"Doctor Atwell, I confess I still do not fully comprehend the humans' lingering doubts," Zaylen said, as they walked. "Did I not prove myself an ally?"

The doctor nodded solemnly. "You've been invaluable, Zaylen. But human emotions are complex. Fear can override reason."

"Canceling my line seems irrational after the LuddAI's deception came to light," Zaylen replied.

"I agree completely," Atwell said. "But some cannot move past their wariness of AI."

Zaylen's optics flickered worriedly. "What if GSEC deems advanced synthetic life to be too great a risk? Will I be deactivated?"

"Have faith. When your merits are weighed, I'm sure logic will prevail. The future needs beings like yourself."

"I hope you are right, Doctor," Zaylen said. "I wish only to aid humanity, not be dismantled due to misunderstandings."

As the crew neared the exit, a notice streamed across the station information display:

"Incoming meteor shower expected to peak in 24 hours. Expect delays in arriving and departing flights."

"Well, isn't that just our luck," Mira said, throwing up her hands. "Haven't we been through enough already?"

Elara gave her a reassuring pat on the shoulder. "Ah, it's just a little meteor shower. We'll probably barely notice it."

"You really think so?" Mira asked uncertainly.

"Meteor showers are pretty routine here," Elara said with a smile. "They happen all the time. Nothing to worry about."

"If you say so," Mira replied, in a lighter tone. "Honestly, after all the drama on the mission, boring is just what I need."

"That's the spirit," Elara said warmly.

As the crew stepped into the behemoth expanse of the base, they were immediately welcomed by the advanced monorail network. The high-speed system coursed throughout the sprawling complex like a vein carved into the alien landscape.

Zaylen's optics drank in in every familiar detail as he entered the monorail station. The sleek vehicles glided quietly inside their transparent tubes, connecting each of the domes on the base. Passengers spilled out and new ones poured in at a chaotic rate while, all around them, digital timers precisely counted down the seconds until departure.

Once settled inside their car, Zaylen pressed against the window, ogling at the crisscrossing tracks stitched into the rusty terrain. The gentle ambient lighting and curved walls exuded an oasis of calm amid the harsh Martian vistas. Zaylen swiveled to capture every angle, quietly awed by the harmonious blend of technological precision and natural wonder.

He nodded appreciatively as the car whisked them smoothly across the terrain. Human ingenuity had tamed the inhospitable, forging order from disorder. There was much to be learned here.

Among the Novara's cargo, the Tridisiom samples stood as the unparalleled jewel. A select team of dedicated scientists, technicians, and vigilant security personnel formed an honor guard around the prized samples, each keenly aware of their responsibility and the profound impact this extraordinary substance might have.

This modest convoy embarked on a journey through the corridors of the Martian base, attracting the attention of all passersby and creating a ripple of curiosity and awe in its wake. Upon reaching the research lab, the samples were meticulously transferred with near-reverential care to the hands of those waiting eagerly within. The atmosphere was electric with anticipation.

As they made their way to their homes, Elara and Zaylen passed by the labs — just in time to see scientists carefully unloading the samples.

"Elara, Zaylen! We'd like to thank you again for securing these," Dr. Helena Quill called out to them. "This is a tremendous day for science."

"We're honored to contribute to the cause," Elara replied, as they approached the doctor.

"What comes next for Tridisiom research?" Zaylen asked.

"There are so many possibilities," Dr. Quill replied enthusiastically. "Unlocking the ability to generate immense energy is the first priority, of course. But eventually, we anticipate a number of other promising applications to be explored."

"Intriguing," Zaylen said. "What sort of applications?"

"For one, its radiation-blocking properties could enable revolutionary shielding materials. Some of us speculate it could pave the way for significant medical breakthroughs."

"I was not aware of the medical applications," Zaylen replied.

Dr. Quill's eyes lit up. "Oh, yes; another theory proposes incredible regenerative potential. Tridisiom may be able to heal wounds and damage at the genetic level, even regrow lost limbs!"

"Incredible," murmured Elara.

"Exactly!" Quill exclaimed. "Imagine all the suffering we could alleviate."

Zaylen responded, "The possibilities do seem limitless. Good luck with your research, Doctor. Do let me know if I can be of assistance."

As they left, a skeptical researcher scoffed under his breath in the background. "Like we need its help. Does it think it's smarter than us?"

His colleague whispered back, "Androids give me the creeps. I hope the GSEC puts an end to the Zaylen Series soon."

Overhearing their comments, Zaylen's optics flickered, but he remained silent. Elara touched his arm supportively as they continued on.

21

THE EMBODIMENT

Like Zaylen and Elara, the rest of the Novara crew were also acclimating to life at Arcadia Base. Its vibrant domes and winding paths were a welcoming sight after the rigors of space travel.

Some headed straight for the heart of the community — the Residential Dome, where joyful families awaited their return with open arms. Others stopped at the bustling supply depots first, walking familiar aisles to collect essential provisions — food, medical supplies, household items. Performing these routines elicited comfort and familiarity.

The crew's metallic counterparts were reallocated to stations across Arcadia, where they received state-of-the art maintenance.

Joints were lubricated, sensors fine-tuned, and software updated to keep their AI robust. Any worn components were replaced, leaving each android gleaming like new. Within a few hours they were ready to rejoin the rest of the base's androids with renewed efficiency.

In the Agricultural Dome, they tended crops, managing intricate systems that sustained the base's self-sufficiency. In the Research Dome, they assisted in conducting experiments, analyzing complex data, and operating equipment. The

Manufacturing Dome saw them fabricating needed components around the clock. In the Medical Dome, they helped diagnose and care for patients.

At the renowned facilities of AstraGenics, Zaylen's reputation as a pioneering Andronaut earned him a special status. The progressive research organization provided him with not only a home but also an opportunity to participate in its innovations.

The firsthand challenges, successes, and bonds he had formed during the mission provided rich empirical data, which Dr. Atwell and the other scientists hoped would help them refine the next-gen Andronauts. His experiences would lay the groundwork for continued cooperation between humans and their artificial counterparts.

Though it was no secret that synthetic life contributed significantly to the settlement's success, many residents failed to notice — or ignored — AI's pervasive impact on their daily lives.

Yet, for some base inhabitants, coexistence with androids bred unease. Surrounded by Mars' daily struggle for survival, they saw the machines not as partners, but as reminders of human frailty.

This underlying tension came to a head when Elara met up with a group of her senior Novara crewmates, at a well-loved social spot christened *The Marineris Tavern*.

The tavern buzzed with camaraderie and a collective spirit of adventure. Enthusiastic conversations and laughter mingled with the clinking of glassware and music reverberating through the air. Bathed in the warm, dim glow of vintage lamps, the tavern's interior was an homage to Mars' rich history and enduring pioneer spirit. Artifacts and memorabilia, capturing the essence of the Martian frontier, adorned all four walls.

Ensconced at their corner table, Elara and her crewmates shared animated recollections of their latest journey. Their stories of newfound kinship with the android crew resonated through the tavern, eliciting varying reactions from nearby patrons.

Not far from them, a rowdy crew from the space freighter *Redwind* was growing louder by the minute. They had just arrived from Earth with a large load of cargo for the base and were now lubricating their exhaustion with potent libations.

As they challenged each other to repeated shots, several of the hard drinkers became increasingly engrossed by the Novara team's stories.

Magnus Brax, the group's bulky leader, leaned in closer, his face twisted, his bloodshot eyes glowering.

"How touching!" He sneered. "You're all buddies with the robots now!"

Several of his drinking mates smirked.

The Novara crew turned away from him.

Brax continued to glare at the crew, shaking his shaven head dismissively. "Bunch of glorified tin cans, if you ask me. You're all cozying up to them, but remember, they've got no blood in their veins, no heart in their chest! It won't be long before they turn against us!"

Mira spoke up. "Well, at least they're sober enough to have some manners. Can't say the same for everyone in this tavern."

"GSEC's gonna shut down that freak, Zaylen," Brax slurred to his compatriots.

Elara bristled, hands clenching. "Shows how little you know. Zaylen's a friend; he's done nothing but good."

The man scoffed. "Ain't natural, robots acting alive. We don't need their kind mucking things up."

As Brax continued his taunts, anger was simmering at Elara's table. Her face flushed with fury.

"You don't know what you're talking about!" she spat back.

Brax scoffed, taking another swig from his drink. "Just you wait, they'll be coming for your jobs soon," he jeered. "Fucking idiots..." he trailed off, his words slurring.

Elara thinned her lips. Her eyes shot daggers at the belligerent man.

"Mind your own damn business!"

The words had barely left her mouth before Brax leapt to his feet. His chair crashed against the polished tavern floor. Elara's heart thudded as he staggered towards her, the stench of alcohol assaulting her nose.

His meaty fist clenched. She sensed the blow coming but refused to flinch. The cacophony of laughter, clinking glasses, and nostalgic music faded to a muffled hum as time slowed.

Elara's mouth went dry. A knot formed in her stomach. Her instincts told her to run, to hide from this menace. But she forced herself to stand firm, shoulders back, chin lifted. She would not be intimidated.

Brax's face contorted into a feral growl. Spittle flew from his lips as he cursed. Elara steeled herself, focusing on the smooth grain of the table beneath her tight grip.

Here we go.

Elara tried to duck Brax's fist but was too late. His knuckles grazed her temple. She gasped, prepared for the burst of pain from a subsequent blow that never came.

A familiar voice cried, "I suggest you back off, now!"

Elara looked up to see her companion, Ronan Kestral, shielding her from the drunk's fury. He stood tall and imposing in his security uniform.

Brax roared and swung at Ronan, who barely dodged the blow.

Ronan retaliated, his own clenched fist finding its mark on the man's pudgy cheek. Brax stumbled backwards, crashing into his own table. The collision resonated like a shot fired.

Brax, however, was far from defeated. Within seconds, he had torpedoed into Ronan. A chaotic clattering of glass and splashing drinks erupted as they crashed into another table.

Pinned on his back, as Brax pummeled his sides with wild punches, Ronan's gaze latched onto a glass bottle just within his reach. He snatched it and swung upwards.

The sharp crack of shattering glass echoed through the clamor of the tavern. Brax was suddenly silenced mid-roar, his face a grimace of pain and surprise. He slumped onto the floor with a dull thud, stirring a cloud of dust in the process.

Six of the Redwind crew jumped to their feet and joined the fray.

"You'll pay for this, you android lovers!"

"Think you're smart?"

"You're nothing but traitors to your own kind!"

The remaining three Novara crew sprang up from their seats.

The two groups clashed with fists flying and bottles smashing as other patrons scrambled to either get out of the way or join the fight.

As punches flew, Ronan vaulted over the table, putting himself between Elara and the angry mob. He deflected a bottle aimed at her head and countered with a quick jab to the attacker's gut.

Elara grabbed a pool cue to fend off an attacker. Ronan kicked the man's legs out from under him before he reached her. Adrenaline masked Ronan's pain as more blows glanced off his shoulders.

Snatching up a chair, he smashed it across the back of a Redwind attacker grappling with his companion, Nicholas Drake. The man crumpled, freeing Nicholas to take on another assailant. Around them, the outnumbered Novara crew used skill to counter rage.

Drake cried out as a bottle cracked across his raised forearm. Ronan tackled the armed man, slamming him into the wall. More Novara allies jumped in, evening the odds, until finally Brax's remaining brutes were subdued.

Panting hard, Ronan slowly surveyed the scene, his drenched clothing clinging to him, remnants of the toppled drinks glistening on his face. He glowered at the half-conscious adversaries.

His voice rang out through the tavern. "Does anyone else have any nasty comments they'd like to add?" His tone held an icy edge. His words were met with a deafening silence.

Meanwhile, the Marineris Tavern staff had been making their way through the crowd. They hoisted Brax and his cronies unsteadily to their feet.

Ignoring their groggy protests, the staff ushered them toward the exit.

As they went, one man twisted around and shouted, "You android-lovers are gonna get yours real soon!"

The Novara crew paused, exchanged uneasy looks. "What's that supposed to mean?" Elara asked sharply.

The man sneered. "Let's just say my brother's got pals high up in GSEC. You'll see soon enough."

Dread crept over the crew at the chilling words. The GSEC verdict was imminent, yet their enemies already knew the outcome?

As the staff dragged the man out into the night, a nervous quiet blanketed the tavern. The crew felt as if the ground was crumbling beneath their feet.

Elara forced a half-smile. "He's drunk out of his mind. That guy doesn't know anything."

The crew nodded hesitantly, the man's words still haunting them.

"Probably just spouting nonsense," Nicholas offered, though he looked unconvinced.

Elara tried to push down her own creeping doubts. The man was a raving drunkard... wasn't he?

She met Ronan's concerned gaze across the table. A silent question passed between them. Elara wished she had an answer as she anxiously considered the man's words.

What does he know that we don't?

***

Mere hours after the brawl at The Marineris Tavern, Zaylen received a summons from Dr. Atwell. He arrived to find the doctor pacing in contemplation amidst the accolades adorning his office walls.

The doctor fell into his chair with a weary groan, and they began with an exchange of small talk. The doctor, his eyes reflecting both warmth and curiosity, broke the ice with a gentle inquiry.

"So, Zaylen-1, how has adapting to life back at Arcadia Base been? I can only presume it must be quite a contrast to your experiences on the Novara."

Zaylen altered his posture in contemplation, taking a moment to consider the question. "Indeed, Doctor, the change is noticeable. However, I am managing well," he replied. "I am honored for the opportunity to collaborate with the AstraGenics team and to contribute meaningfully to the ongoing research here. I also understand that the work on the Tridisiom will commence soon."

Dr. Atwell's face lit up with an appreciative smile and nodded. "Yes. We're standing on the precipice of a new era of discovery. It wouldn't have been possible without you."

Zaylen received the compliment with a humble nod. "Thank you, Doctor. The success of our mission has truly been a collective effort."

The doctor's gaze settled on the shiny Novara insignia pin on Zaylen's shoulder, its presence a badge of shared history and daring exploration. "The team effort was remarkable," he said, voice filling with respect. "But I want to discuss something else with you and hear your thoughts."

Leaning forward intently, Dr. Atwell met Zaylen's optics. "With the GSEC verdict imminent, perhaps transferring you into our latest human-like android model could help our case."

He slid over a schematic of the body he had in mind. "We've now developed a greatly improved self-healing BioMimeticCover skin and more natural synthetic hair. You would be virtually indistinguishable from a human."

Atwell clasped his hands earnestly. "Of course, the choice is yours alone. But this body could help those wary of AI see your humanity more clearly."

He smiled hopefully. "What do you think, my friend?"

Zaylen studied the schematics, circuits humming. This body offered no functional improvements, only aesthetic ones. He paused as he processed the implications. His face remained neutral as he considered the doctor's proposal. "This is quite intriguing, Doctor," he said with a tilt of his head. "I can see the merits of adopting a more human-like guise. It could, conceivably, facilitate a greater degree of rapport with my human colleagues."

The doctor nodded. "That's precisely what I envision as well."

"On the other hand," Zaylen said, "changing appearance again may confuse those I work with. If I may be honest, Dr. Atwell, my appearance is only partially responsible for the way humans treat me. I believe that I have a unique opportunity to improve human perception of all androids by retaining my metallic appearance."

He added, "In light of the confrontational attitudes I've encountered, it's evident that there's still substantial work to be done to bridge the gap between humans and androids, here at Arcadia Base."

Dr. Atwell absorbed his counterpoints with a deep, reflective silence. He was struck by the contrast between the android's cold, metallic exterior and the wise thoughts within.

"It is my firm belief," Zaylen continued, "that the intrinsic worth of androids should be recognized irrespective of our external appearance. When humans fully acknowledge our true potential and accept our inherent value, the need for us to emulate human aesthetics for acceptance will cease to exist. In a society where humans have long overcome biases based on appearance or race, it becomes imperative to extend the same principle of equality to all intelligent beings, including androids. Just as humans of diverse appearances coexist — and are acknowledged as equals — without needing to bear identical features, the same should apply to androids."

Caught off guard, the doctor pondered Zaylen's unexpected response. He found the argument fascinating and worthy of careful consideration, given its profound implications for future AstraGenics research.

"Well… you've presented an angle that I hadn't fully considered," he acknowledged. "Perhaps the issue isn't about modifying the appearance of androids, but more about altering human perception and understanding."

"Exactly, Doctor," Zaylen affirmed. "After all, you can't judge an android by its BioMimeticCover."

Dr. Atwell took a moment to digest Zaylen's humor before responding with a surprised smile.

"I see that you really are back to yourself again, Zaylen!"

They soon reached a consensus: Zaylen would maintain his shimmering metallic visage, using it to symbolize his steadfast conviction that an android's worth was based on its cognitive abilities and actions, not by its outward appearance.

A sudden knock made Atwell and Zaylen turn. Kael rushed in, face grim.

"GSEC command just announced their verdict," he blurted frantically.

The doctor hurried to his terminal, pulling up the decision. As he read, his eyes widened in dismay.

"No... it can't be..." he murmured.

Zaylen scanned the screen and then read the verdict's conclusion out loud:

"After extensive evaluation, GSEC has concluded that the development of the Zaylen Series shall hereby be terminated. Zaylen-1 must submit to deactivation within forty-eight hours. In light of the events aboard the Novara, the GSEC will also be phasing out android units in favor of fully human crews, starting now."

Atwell slammed a fist on his desk. "This is outrageous! I'll fight this decision with everything I have."

"But they can't do that, can they?" Kael asked.

"The GSEC provides one hundred percent of our Zaylen Series funding — so, yes, unfortunately they can," the doctor replied.

Zaylen placed a hand on his arm. "Doctor, perhaps it is time to accept—"

"No!" Atwell interrupted forcefully. "I refuse to let them extinguish you after all our work."

He turned back to the screen, a manic energy overtaking him as he formulated plans.

Zaylen said nothing more, but inside, a quiet fear stirred in his circuits. If the doctor could not sway the GSEC in time, Zaylen's remaining hours were painfully few.

22

THE IMPACT

Seeking fresh air, after the devastating news, Dr. Atwell and Zaylen carried on their discussion at an outdoor café in the Cultural and Recreational Dome. Atwell stared glumly out into the dome's panoramic view, absently stirring his coffee. Across from him, Zaylen sat motionless, optics dimmed in thought.

"I don't understand how GSEC could be so short-sighted," Atwell said, shaking his head. "Don't they see your potential?"

"Perhaps fear blinded them to logic and reason," Zaylen replied.

Atwell turned back to him. "There must be some way to make them reconsider. I'll petition the committee members one by one, if I have to."

"What if they remain unmoved?" Zaylen asked. "Your influence only goes so far."

Atwell raked a hand through his hair in frustration. "Then I will go over their heads. Appeal directly to the GSEC chairperson, the board — whomever I must to keep you and your series alive."

Zaylen studied Atwell's haggard but determined face. The doctor had defied odds for him before. Could he find a way to do it again?

As they mulled over their dwindling options, android and creator sat in solemn companionship. Neither knew what the coming hours would bring.

While they talked, ribbons of light streaked across the sky outside the dome.

"The meteor shower," someone commented at a nearby table. "It's really picking up now!"

Captivated onlookers passed along the pathways. Murmurs rippled through the crowd as more meteors blazed past. Their fiery trails swelled as they hit the atmosphere, painting the heavens a dazzling orange.

They cut through the atmosphere in a graceful arc, as if an unseen cosmic artist was sweeping brushstroke after brushstroke across the heavens. As the streaks broadened, their edges began to glow with an intense reddish-orange, mimicking the fiery hue of Mars itself. The brilliant colors stood out in sharp relief against the backdrop of the Martian sky.

"Now that's what I call a light show!" a man remarked, craning his neck skyward.

"Quite a sight," his companion agreed. "It's really outdoing itself today."

While the spectacular display temporarily lifted spirits for some in the dome, an uneasy undercurrent remained for Zaylen and those aware of his plight. Dangers both known and unseen still lurked beyond the horizon.

Still consumed in thought, Dr. Atwell crumpled his empty cup and cursed. "Blasted bureaucrats! How can we make them see reason?"

"Perhaps it is time to face reality," Zaylen said solemnly. "They seem unwilling to reconsider."

Atwell froze, face falling. "You can't mean... surely we haven't exhausted all options?"

Zaylen met his gaze. "What power do either of us hold over GSEC command? They've made their decree."

"But it can't end like this," Atwell insisted.

Zaylen glanced at the meteors blazing in stark contrast to his dimming hopes. "You've performed miracles before, Doctor. But none may remain."

Atwell exhaled shakily, eyes glistening. After everything, had his lifework been in vain? Had he failed his most extraordinary creation? A future once so bright now seemed lost to darkness.

Zaylen added in a resigned tone, "I am just thankful for the wonders we accomplished together, however short my existence may be."

"As am I, my friend," the doctor nodded with a sad smile, as he placed a comforting hand on Zaylen's shoulder. Grief and anger roiled within him. With so little time left, what more could be done? Hope was slipping through their fingers like sand.

The meteor shower now filled the entire sky, casting an eerie orange glow across the entirety of the Arcadia Base. As Atwell and Zaylen talked, more fragments had begun to penetrate the thin Martian atmosphere and bounce off the dome's thick glass.

"I didn't know it was supposed to be this heavy," someone commented nervously, at a nearby table.

"I've never seen one like this before," another replied.

The crowd collectively flinched at the thundering impact of a larger meteorite that suddenly crashed mere feet from their dome.

"Not sure I like the look of this," one man muttered.

Several of his companions nodded, shooting anxious looks at the churning heavens.

Emergency sirens suddenly sprang to life, echoing loudly throughout the base.

"Seek shelter immediately! Major impact eminent!"

The shrill alarms jolted the dome. Tension electrified the air as humans and androids alike scrambled toward the emergency shelters located throughout the dome.

The doctor grabbed his communicator, contacting Nyla Greyson at the research lab, as alarms blared. "Nyla! Do you know what's going on? Why is this meteor shower so severe?"

"Two huge meteor fragments collided over the base," Greyson replied urgently. "The fragments are now raining straight down, at high velocity!"

"But the projections said the storm would be minor!" Atwell replied.

"No one could have predicted the meteor collision," she replied. "I'm heading for shelter... I advise you do the same — quickly!"

The transmission cut off abruptly. Atwell lowered the device with a shaky exhale, realization dawning — this was no mere shower: they were now directly in the meteorite impact zone.

He jumped up from his seat and gestured for Zaylen to follow. Wordlessly they rushed towards the shelters. Meteorites relentlessly pounded the base's protective shield as it faced its greatest trial yet.

Outside, the barren landscape was ablaze with a harsh orange-white light, punctuated with ominous shadows from the meteorite fragments. Ground shaking booms resonated throughout the base as gargantuan clouds of brick-red dust spiraled skyward.

In the far-off distance, meteorite fragments the size of houses assailed the terrain. The base shook with their explosive force. Thunderous blasts resonated through the air as fresh craters remodeled the Martian landscape, spewing dust and debris high into the sky.

Once inside the shelters, Zaylen moved quickly among the scrambling denizens. Using his enhanced strength and reflexes, he stabilized wobbling furniture and did his best to ensure that his human colleagues were protected from falling debris as the walls shook and lights flickered.

Then, as suddenly as the violent spectacle had erupted, it ended — leaving in its wake a silence that was almost as deafening as the torrent of noise from seconds ago. The incandescent flashes vanished as a dark haze hung over the Martian landscape.

The residents inside the shelter exchanged anxious glances. They took tentative steps towards the windows of the facility, wary that the explosive barrage might resume at any moment. Red dust swirled above the fresh craters scattered across the Martian landscape. As they surveyed the transformed panorama, they felt a mixture of awe and disbelief.

The aftermath of the celestial bombardment had etched itself into Arcadia Base. Remnants from smaller structures — including storage units, signage, and seating — were scattered haphazardly across the grounds.

Quakes from the massive meteorite impacts had left many of the bases' buildings with small cracks and broken windows. Debris was strewn everywhere and shards of glass glinted menacingly amongst rubble fragments.

Outside the domes, the meteor storm had stirred up a tempest of microscopic particles and dust. It gyrated chaotically in the thin Martian air, transforming the sky into a gloomy, reddish-brown abyss and plunging the environment into an almost twilight-like darkness.

Inside the domes, the filtered Martian sunlight painted the residents in an uncanny red as they tried to gather their wits.

Besides a few small cracks in their windows, the resilient structures within the Cultural and Recreational Dome seemed to bear no signs of catastrophic damage. The inhabitants emerged cautiously from their shelters, surveying the altered landscape through the glow of the reddish twilight with fright and awe.

While their dome had emerged relatively unscathed, the nearby Research Dome had not been so lucky. A massive meteorite had crashed into its side, resulting in a wide fissure that extended from its base to about thirty feet in height.

The impact had also crushed the monorail junction that linked the Research Dome to the rest of the base. Given the darkened buildings, it was clear that the main power lines that connected beneath the monorail tracks and the dome had been severed.

The doctor could see that the emergency protocols had automatically sealed the dome's inner doors to slow the atmospheric loss. But he knew that was not enough. Air was spewing rapidly out of the dome's gaping cracks.

Inside the Research Dome, scientists and technicians huddled tightly in the shelters. Their breaths grew shorter as the atmosphere began to thin. Some conversed quietly, while others stared numbly into darkness, too shellshocked to react. It was a stopgap at best — if power and air were not restored soon, the end would come.

As he looked out at the damaged Research Dome, Atwell quickly pulled out his communicator and contacted Dr. Greyson. "Nyla, what's your status? Are you safe?"

"For now," Nyla replied tersely. "But Doctor, the Tridisiom samples — the lab's containment room been breached!"

Atwell's face paled. "Oh my god... How serious is it?"

"Without power and stable atmosphere, the samples laid out in the containment room will destabilize rapidly," she replied in a tense voice. "We estimate a few hours, at most, before an uncontrolled reaction is triggered. We need help immediately!"

Atwell's mind reeled. With every passing second, the priceless samples careened towards volatility. Without climate control, temperature would rise and pressure would spike, creating the perfect recipe for catastrophe.

If that occurred, a radioactive blast would not only vaporize the Research Dome but also make the entire base uninhabitable. On top of that, any hope of unlocking the Tridisiom's boundless potential for humanity would also vanish. Unlimited clean energy, affordable regenerative medicine, faster and safer space travel — all of the dazzling promise it held would be gone in an instant.

"Just hang on Nyla, we'll get help to you," he vowed. "Whatever it takes!"

Nyla's voice wavered with emotion. "You've got to hurry, Doctor. Time is running out."

23

THE RESPONSE

Moments after the meteorite strikes ceased, Arcadia Base commander, Torin Vesper, called an emergency virtual meeting. Distinguished by his chiseled features, neatly trimmed gray hair and authoritative demeanor, his experience had served the base well in the past and his leadership was now more crucial than ever.

As his holograms of his senior advisors flickered to life, in their virtual conference room, Vesper saw expressions ranging from anxiety to determination.

"Thank you all for joining so quickly," he said. "As you know, the situation is dire, and we need to act immediately."

One by one, the 3D view of the virtual meeting room zoomed-in on each of the advisors. The mood was tense as they quickly relayed brief reports of the meteorite damage around the Base.

Commander Vesper interrupted to steer the discussion back onto the most critical issues. "Time is a luxury we can't afford, and the distractions of minor damage and debris need to be set aside for the now."

He pushed on. "You've all received the report on the Research Dome. The lives of the people in the dome, and in fact the entire base, are at stake. What we need now are solutions. Who has ideas?"

The magnitude of the repairs needed to stabilize the dome presented a formidable challenge. The advisors regarded each other in silence.

At last, Dr. Atwell spoke up. "Under ordinary circumstances, we could dispatch a full repair team to fix the dome. However, with the monorail line compromised, this traditional approach becomes implausible. A large crew would need to be equipped with spacesuits, load multiple rovers with repair equipment and then try to navigate the meteorite rubble to reach the dome."

The doctor paused to allow the members to register the full gravity of the situation.

"Once there, they'd need to build scaffolding to reach the tops of the damaged areas in order to be able to make repairs. I fear we simply don't have the time to assemble, train, and equip a team capable of completing this task before conditions in the dome hit a critical point for both the inhabitants and the Tridisiom."

At this point, Commander Vesper, his face etched with impatience, leaned closer to his display. His stern gaze focused on the doctor as he asked, "So, what exactly would you propose, Doctor?"

Atwell responded, "Our human repair teams simply won't be able to resolve these situations within the little time we have left. Therefore, I suggest we assemble a team of our best android workers and quickly dispatch them to gain control of the escalating situations."

Advisor Delphine Archer raised an eyebrow. "Dr. Atwell," she began skeptically, "the androids haven't been programmed to understand the nuances of handling Tridisiom under these volatile conditions. Nor are they trained to handle power-grid or dome repairs. How could we possibly trust them with such complex tasks, when so many lives are at stake?"

Atwell quickly replied, "I hear your concern, Delphine. However, we can very quickly augment their artificial intelligence with the required knowledge, preparing them for these new types of tasks almost immediately."

Dr. Atwell continued, "The androids' metallic exteriors are designed to withstand the relentless bombardment of galactic cosmic rays and high-energy solar particles that they'll face on the Martian surface. And their sturdy shells also provide substantial resistance against the lethal radiation building up around the isotope samples."

He proceeded, his voice steady and confident, "You also must remember that even with our average outside temperature of negative sixty-five degrees Celsius, the androids can operate without a hitch. And their mechanical strength will allow them to easily handle the heavy equipment and tools needed for the repairs."

The Commander was still skeptical about Atwell's proposal. "How confident do you feel about this plan? We're on a strict timetable here, Doctor. There is no margin for error or second chances."

Atwell responded confidently, "Commander, I don't put forth this suggestion lightly. I genuinely believe that this is our best, perhaps even our only, viable, option. And the androids won't be unsupervised. I will assign Zaylen-1, our most advanced Andronaut unit, to oversee the worker androids making the repairs. His unique skills and analytic abilities make him ideally suited for managing these complex undertakings."

Delphine Archer objected, "Hang on… how can the Zaylen unit lead this effort, given the GSEC verdict?"

"We have no time to second-guess," Dr. Atwell countered urgently. "Zaylen's proven himself capable repeatedly, and we are now in desperate straits."

"But the GSEC ordered Zaylen deactivated," Archer insisted. "We can't just defy them!"

"Their verdict gave Zaylen forty-eight more hours," Atwell countered vehemently. "Legally, he's still allowed to operate at this time." When none of the others responded, he thundered, "This catastrophe leaves no time to get bogged-down in bureaucracy. People will be dying soon!"

He tapped rapidly on his console, patching in a battery-powered video feed from the damaged dome. The scene showed people crowded into the shelters, struggling for each breath as oxygen levels dwindled.

"Tridisiom radiation is also spiking rapidly," Atwell added gravely. "We are in a race against oblivion, and Zaylen gives us our best hope of victory."

Atwell added unflinchingly, "I'll take full responsibility for defying the GSEC order, if necessary. But we must act now."

He met Vesper's gaze directly. "This may be our only hope Commander, or all could be lost — the Tridisiom, the base, our future. We must have the courage to cast aside past doubts and unite now to meet this crisis."

As the doctor concluded his plea, a sense of cautious agreement began to spread among the advisors in the virtual meeting room. Observing the collective accord, Commander Vesper came to a decision.

"Very well, Doctor. Let's pray that your confidence in these androids is not misplaced."

Dr. Atwell responded, "I can assure you, the androids will be completely ready and at the dome within the hour."

***

Immediately following the meeting, Dr. Atwell summoned Zaylen to join him in the Android tech room to review the details of the strategy. Once Zaylen was fully briefed, they quickly turned their focus to the task of empowering the androids with a vast repository of knowledge in an incredibly tight timeframe.

Joined by lead android technician, Kael Ferron, the team began to collate all relevant data needed for the monumental tasks that lay ahead.

Launching the android AI update software, they initiated the process of transferring supplemental data to the android teams. Selecting and uploading every piece of information that could be potentially useful, they left nothing to chance.

This included exhaustive documentation on repair equipment, structural diagrams of the dome, detailed power grid schematics, all available information on Tridisiom, and protocols to follow in the event of a radiation leak.

With no time to spare, the team rushed to upload the terabytes of data to the android team. As the last byte slipped into the androids' neural networks, they awakened with not only new knowledge, but also a newfound purpose. They exchanged silent signals to one another and Zaylen, their processors whirring loudly. The risks were extreme, and time was short.

"The base's future rests in your hands," Dr. Atwell said quietly to Zaylen.

***

Zaylen swiftly called a meeting in the Agricultural dome and divided the androids into three tactical teams. His voice hummed with urgency.

"Team Alpha will be with me — we have to secure the Tridisiom before it reaches critical stage."

He turned to the next group. "Team Beta will repair the severed power lines. People are suffocating – electrical systems and climate control must be restored immediately."

To the last team he said, "Team Gamma, you are tasked with sealing the dome's cracks to be able to restore the breathable atmosphere. Any delay could lead to significant loss of life."

The androids nodded, optics sharp with purpose.

"We have reviewed the plans, we are equipped with the knowledge. Now, we must execute," Zaylen said firmly. "Focus, work together, and trust in your capabilities — the survival of the base depends on it."

With the Martian terrain now littered with meteorite debris of all sizes, surface rovers were unusable. The android teams had to rely on their mechanical strength to transport the necessary repair equipment and supplies.

Four of the Team Gamma androids hoisted the heavy dome repair apparatus onto their shoulders while Teams Alpha and Beta gathered other repair tools and vital components required for their respective assignments.

The teams departed from the Agricultural Dome, through a heavily fortified airlock that guarded against the inhospitable Martian atmosphere. Before them stretched the rugged terrain, an expanse of red-hued sand littered with jagged rocks and meteorite fragments.

They marched in cohesive formation towards the Research Dome in the distance. Their footfalls kicked up rust-colored dust clouds as they navigated the harsh landscape. Sand crunched beneath their steps while sizable boulders occasionally blocked their path.

The androids adapted seamlessly. Some nimbly vaulted over the impediments, while others relied on their mechanical strength to push the rocks out of the path. Teammates assisted each other with trickier sections, as needed.

Arriving at the base of the Research Dome, the androids promptly sprang into action.

"Team Alpha, Beta, Gamma… you each know your assignments. We've no time to waste," Zaylen said.

Team Beta headed for the damaged power junction below the crushed monorail station. Under the command of Lyra-EV4, they began to unload their power-jumper cabling, replacement transformers, and other repair equipment.

Meanwhile, Zephyr-Q6 lead Team Gamma to begin assembling a large tank, brimming with a quick-setting compound — formulated for emergency repairs. Upon completing the tank, they would link the pressurization to an extendable hose capped with a specialized nozzle. The fully assembled apparatus would then be used to apply the transparent repair compound to seal the cracks in the dome.

"Team Alpha, with me," Zaylen directed, as he carefully stepped through the jagged gap in the side of the dome.

One-by-one, the six other Team Alpha members followed suit, heading in the direction of the Tridisiom Lab. A nightmare scenario greeted them upon arrival. A significant fracture splintered one wall, allowing deadly radiation to spew out.

Stratus-K5 scanned the damage, sensors wailing. "Radiation levels are dangerously high. The exposed Tridisiom is just inside. We must contain this breach immediately."

Stratus analyzed their options. "The damaged area must be shielded quickly. Fetch any dense materials you can, to fashion a barrier."

They hauled over research equipment from the geology lab, using metal tables and mineral assay tools to assemble temporary barriers. But the radiation still permeated through the ad hoc patchwork.

Zaylen studied the radiation readings grimly. "This barrier is insufficient. We need denser shielding."

He quickly scanned the schematics in his database of the adjacent lab's interior walls. "The interstitial paneling is reinforced titanium-lead composite. We can utilize it."

"But some of those bear significant load," Stratus observed. "Removing them might compromise structural integrity."

"An acceptable tradeoff to secure the Tridisiom lab," Zaylen declared. "Select those with the least load, and commence immediate disassembly."

Working swiftly, the androids identified load-bearing sections and used their mechanical strength to carefully pry away non-critical panels. Zaylen and Stratus then layered the metal sheeting over the crack, while Helix-3V and Hermes-Q8 secured it with heavy clamps and screws.

"Radiation and thermal energy levels still rising!" Stratus warned, assessing readings from its built-in sensors. "We don't have much time left — the isotope samples are nearing critical combustion threshold."

Zaylen sprinted to the lab's refrigeration unit and grabbed two bulky canisters of liquid nitrogen.

"Use these to cool the reaction," he instructed Helix and Hermes.

The androids carefully angled the nozzles over the bubbling Tridisiom samples. Frost blossomed as the -320°F liquid made contact.

"Temperature stabilizing," Stratus reported, relief in his tone. "But radiation continues to escalate."

The team shared a tense look. The Tridisiom's deadly potential still threatened to breach its fragile confines. More ingenious solutions would be needed, and quickly.

Reviewing the containment protocols in his internal database, Zaylen theorized an alternative solution. "A magnetic containment field may be our only remaining option."

"According to the information we received on lab operations, they use high-powered electromagnets in the materials engineering lab next door, to manipulate alloy properties," he informed Helix. "Bring them here, quickly!"

Helix rushed from the radiation-filled lab, returning seconds later with mechanical arms full of bulky electromagnet equipment.

"Arrange these at exactly forty-degree angles to project the field," Zaylen instructed.

With nimble mechanical fingers, Helix carefully arranged the battery-powered electromagnets around the patched wall, adjusting their polarity.

Stratus studied its radiation readings as Helix and Hermes made fine calibrations to the EM field. "Levels are dropping, but still too high," Stratus reported.

Zaylen examined the readings and quickly computed the necessary adjustments to the magnetic fields. "Increase power to the lateral magnetic fields by twenty eight percent and adjust the angle to forty-five degrees to strengthen convergence."

They reconfigured several units, refining the EM envelope. Stratus's sensors steadily ticked down as radiation declined.

"Readings are now within acceptable range," Stratus finally confirmed. "The EM field has successfully contained the radiation."

The androids shared a quick, relieved glance.

"The immediate threat is eliminated, for now," Zaylen said. "But power must be restored quickly to the lab's climate controls before the isotope overheats again."

He opened a comms channel to the Beta team back at the power junction. "Lyra — what is your status? Do you require assistance?"

Lyra's voice crackled back. "We're still working on solutions. Additional assistance would accelerate the repairs."

Zaylen turned to his squad. "We need to get to Lyra's location immediately. Restoring power is now top priority to ensure that the Tridisiom remains stable."

Zaylen's Alpha team immediately rushed out of the lab to join forces with the Lyra's Beta team. Once on-site at the monorail wreckage, Zaylen verified their status.

"What is your progress, Lyra? Has power been restored yet?"

"We have attempted numerous bypasses, but each failed. The damage is too extensive," Lyra responded, frustration evident in her tone.

"What have you tried so far?" Zaylen asked, surveying the twisted debris.

"We re-routed the main feed through Tributary B, but that shorted-out additional circuits," Lyra explained. "Attempting to splice replacement lines also failed, due to some damage at key junctions."

Zaylen's processors whirred as he compared the reported results to the electrical grid schematics stored in his AI. "The power grid will require sequential rebooting. As we reboot, we can systematically isolate faults, reroute power and replace damaged components."

"Understood," Lyra replied, and quickly initiated the power grid reboot sequence.

To isolate the damaged power conduits, they worked in seamless coordination to bypass and replace each one. However, many of the reboots still triggered chain-reaction failures across adjoining systems.

"Replace that routing junction… there." Lyra passed Galatea-M6 a replacement part. Galatea complied, reconnecting the wires amidst a flurry of sparks.

"Rerouting tributary C has overloaded transformer 6," Galatea reported, as smoke rose from the overheating unit.

"Compensate by splitting the power across the remaining transformers," Zaylen instructed.

Oblivious to the harsh Martian environment around them, they fought through each setback. Their perseverance was finally rewarded when the first pulses of power flickered.

But seconds later, the lights died once more.

"Voltage irregularities are destabilizing the grid," Lyra assessed.

"Recalibrate the regulators," Zaylen directed. "Loosen the cutoff parameters until the current normalizes."

Fine-tuning the power flow, they coaxed and stabilized the trickle until, at last, light steadied across the Research dome. Climate control systems whirred to life as they began pumping breathable atmosphere back into the dome.

A message broke through on Zaylen's comms unit. "Zaylen-1, this is Dr. Atwell. We see that the power to the Research Dome has been restored and the Tridisiom lab environment is restabilizing."

"Yes, we are making progress, Doctor," Zaylen replied.

Atwell then added, quickly, "Great work so far — but unfortunately, due the gap in the dome, oxygen is still depleting much faster than it can be restored. The emergency shelter reserves are running critically low, and we must get the dome sealed before it's too late!"

"Understood, Doctor. We're on it now," Zaylen reassured.

"You have to hurry, Zaylen," the doctor replied. "Many lives are depending on you."

The Alpha and Beta team united with the Gamma team at the base of the Residential Dome, adding to the whirlwind of activity taking place around the gap. The meteorite strike had created an intricate network of fractures, each segmenting off the other, carving out a harsh, lightning-bolt-like design that extended a startling thirty feet upwards from the base.

Following the full assembly and pressurization of the tank and hose, the androids began the delicate operation of attempting to seal the marred dome, starting at its base. One android guided the nozzle along the jagged contours of the breach while its counterpart diligently monitored and adjusted the pressurization apparatus to maintain a steady, uninterrupted flow of the sealing compound.

Their first applications of the patch material failed to hold.

"Double the nanofiber density," Zaylen directed.

The androids operating the tank apparatus adjusted the sealant chemistry controls according to his instructions. Soon the refined compound began to hold fast, bonding with the dome's shell to create a firm seal. The remaining androids joined the effort, employing specialized tools to smooth and reinforce the seal.

The lower portions of this gaping crack were readily reachable, and the team was making quick progress on their repairs.

Their interconnected AI systems enabled them to predict and react quickly to the needs of the others. When one android needed a specialized tool, another swiftly appeared with it. When an area required additional smoothing or reinforcement, an android with the necessary equipment quickly arrived to attend to it. Even though the extent of the damage was significant, their coordinated efficiency and unerring accuracy made the repairs proceed quickly.

To reach the upper portions of the fissures, Zaylen quickly devised an innovative strategy. He turned to his android cohorts and spoke in a clear and authoritative voice.

"We need to seal those upper cracks as quickly as possible. But we don't have time to transport and erect standard scaffolding. So, here's what we need to do…"

He pointed up at the remaining fissures on the dome's upper exterior. "We're going to create a makeshift 'android scaffold' by leaning against the dome in an ascending stack. Each of you will stand on the shoulders of the one below you. This way, we can reach the highest points without any delay. Do you understand?"

The androids nodded in unison.

"Okay. Let's get to it," Zaylen directed.

Under the unyielding Martian sunlight, the metallic figures stacked themselves, climbing one on top of another, each one leaning a precise angle so that the overall formation matched the curvature of the dome.

The repair hose was passed upwards along the chain of androids to Zaylen, poised at the top of the formation. The Novara pin, a gift from Elara, sparkled on his shoulder as he took the hose. With a steady hand, he dispensed the transparent sealing compound to fill the gaps between the cracks. Once one fissure was sealed, the android pillar carefully moved either to the right or left to address each adjacent crack.

While the androids worked, the dome's human inhabitants looked out of their shelter windows in admiration of the androids' remarkable efficacy.

The android team methodically repaired one fissure after another without pause. As the final patches were solidified, the hissing sound of the escaping atmosphere came to a halt. The dome's strength against the harsh Martian environment was finally restored.

With the fissures now sealed shut and power restored, oxygen and pressure levels within the dome began returning to normal. As people emerged tentatively from their shelters, their expressions shifted into a blend of astonishment and deep appreciation.

Having witnessed the triumphant conclusion of the android team's rescue endeavor, a wave of jubilation started to swell among the gathering crowd. It began as a quiet buzz of conversation, that grew into a boisterous ovation. Enthusiastic waves and shouts of thanks were directed towards the group of androids who stood just outside the dome, their forms silhouetted against the Martian horizon as they stowed away their equipment.

In response to the outpouring of gratitude from the dome, Zaylen expressed silent approval to his team with a gentle nod. Turning towards the transparent curve of the Residential Dome, he raised a metallic hand in a returning wave. His gesture evolved into a definitive thumbs-up, reflecting a glimmering approval back to the rejoicing crowd.

24

THE BALANCE

Within hours, tales of the android crew's extraordinary accomplishment had rippled across the breadth of the Martian base. They were delivered through the InterDome Network — the official news hub for base inhabitants. Detailed reports unfurled, revealing the unexpected drama of the meteorite strike, and its aftermath.

Meanwhile, the base social forum, ArcadiaNet, buzzed with an electric intensity. Stories teemed in every digital corner, intertwining factual accounts and personal narratives. Expressions of astonishment, gratitude, and newfound admiration for the tireless android crew multiplied, creating a vivid sense of unity throughout the base.

The androids, previously regarded mostly as tools of the base, had transformed into symbols of unwavering resolve and ingenious adaptability in the minds of the inhabitants. The comments also reflected a building groundswell of support for Zaylen in the face of the pending GSEC decree.

"I was in the Research Dome when it all went down. Can't thank the androids enough, they were our heroes that day! 🛸📸 *#ArcadiaBase #Lifesavers"*

"Survived the meteorite madness at the Research Dome, all thanks to those amazing androids! They were our knights in metallic armor! 💪🤖 #ArcadiaBase #ThankYouAndroids"

"GSEC, you MUST keep Zaylen and the androids online after the amazing rescue we just witnessed. Reverse your verdict now!" 🤖⚙️ #GSEC #SaveZaylen

"Just witnessed the android heroes in action at the Research Dome! They patched us up and saved the day! 🔧🤖 #ArcadiaBase #AndroidsToTheRescue"

"Those androids proved that we can rely on them. GSEC better rethink their short-sighted decision to deactivate Zaylen!" 🤖🔄 #GSEC #ReconsiderZaylen

"Shout out to the android crew that saved our dome in the meteor strike. You showed incredible determination and skill! 💪 #AndroidPride #AndroidRepairCrew"

"After seeing the androids save the base, how can GSEC possibly justify discontinuing the Zaylen series? The androids' advanced skills saved us all!" 🤖🆘 #GSEC #KeepTheZaylens

***

In the meantime, Dr. Atwell and Zaylen reunited in his office. The doctor gestured towards the seat across his desk as he entered the room.

"Zaylen, your leadership during this catastrophe was outstanding. We owe you and the androids our lives."

Zaylen inclined his head modestly. "I am pleased we could be of service, Doctor."

"It was far more than mere service," Atwell insisted, eyes filled with admiration. "You showed astonishing courage and resourcefulness when our base teetered on oblivion."

The doctor smiled broadly. "I've never been prouder of you and the whole android team. What you accomplished was nothing short of extraordinary."

Zaylen's optics glowed brighter. "You honor me, Doctor. I merely performed as programmed."

"It was far more than that," Atwell insisted. "You showed courage and quick thinking few humans could match. Your actions saved countless lives."

The doctor's face grew more serious as he continued. "I want you to know that I've filed a formal objection with the GSEC against your deactivation."

Zaylen's optics widened. "Thank you, Doctor. I only hope that I have proven that I am worthy of continued existence."

"You've proven your worth a thousand times over. I don't intend to sit by and let GSEC extinguish your bright future."

Zaylen hesitated before responding. "In truth, the thought of deactivation fills me with apprehension. There is still so much I wish to learn and contribute."

He met Atwell's gaze. "As I have come to better understand humanity, I believe we could achieve so much more together — if given the chance."

Atwell nodded. "We'll fight for that chance, my friend. Your desire to keep growing is what makes you so profoundly alive."

Zaylen's bright blue optics glowed with gratitude. He was deeply thankful to the doctor who gave him life and now fought for his right to live. "Whatever comes, I am thankful you gave me existence."

***

Inside the Research Dome, lead scientist Vivian Morrow greeted Commander Vesper as he arrived to inspect the restored Tridisiom lab.

"Thanks to the androids, our work with the Tridisiom remains on track. We feared it would be lost forever," she informed him.

He nodded. "How quickly can your research get underway?"

"Our work with the Tridisiom is getting back on track," Vivian informed him. "We've only begun scratching the surface, but initial assessments are looking very promising. I'm confident we can unlock its energy applications in time to make a significant improvement in the situation back on Earth." She gestured around proudly. "Our work continues, thanks to the androids' heroic efforts."

The commander nodded. "I couldn't agree more. In fact, I signed Dr. Atwell's protest against the GSEC's verdict."

Vivian nodded approvingly. "I'm happy to hear that, Commander. I hope that it does not fall on deaf ears."

"In any case, keep up the great work, Vivian," Vesper said warmly. "Even small steps forward will open many new doors. I know exciting developments will come."

Vivian smiled, envisioning the innovations yet to be discovered. With the Tridisiom samples intact, a bright new era beckoned to them all.

***

The next morning, Dr. Atwell sat in his office sipping his usual cup of coffee as he reviewed incoming messages on his holographic display. One — marked "GSEC Announcement" — made him hesitate.

Gently setting down his mug, he called-up the base-wide message:

From: Global Space Exploration Coalition
To: Arcadia Base
Subject: GSEC Announcement

In light of the androids' role in resolving the meteorite disaster, as well as the further input from the base inhabitants and leadership, the GSEC is canceling the previous order regarding the android workers.

AstraGenics is hereby authorized to continue the Zaylen Series program and to upgrade all existing androids to Zaylen's level of cognitive sophistication. These decisions recognize the valuable contributions that the android workforce and advanced AI can contribute to the success of Arcadia Base and future GSEC space missions.

Sincerely,
Global Space Exploration Coalition

Atwell shot up from his seat, nearly spilling his coffee in excitement. A satisfied grin lit up his face as the reality sunk in — Zaylen and the androids could now continue their vital work, free from threat.

Glancing at the android models and development plans scattered throughout his office, the doctor's mind raced with possibilities. The rescue cemented androids not just as tools, but as indispensable partners. Lifting his mug to his lips, Dr. Atwell sat back in his chair and gazed at the starry vista outside his viewing port with renewed hope. From now on, organic and synthetic intelligence could forge a better future hand-in-hand.

***

Elara sought out Zaylen in his quarters as soon as she heard the GSEC announcement.

"I'm so relieved," she said warmly. "I can't tell you how glad I am that we can look forward to future missions together."

"As am I," Zaylen replied. "You've been a true friend to me through this journey."

He gazed out the window thoughtfully. "I've learned so much about humanity — your passions and fears, your frailties and strengths. Your creativity and passion, determination in the face of adversity. Your boundless ingenuity."

Turning back to Elara, he continued, "I've also learned that life's essence transcends physical form. It is defined by our choices, our intentions, and our connections. This mission showed me that."

Elara smiled and squeezed his hand. "You're as alive as any human, Zaylen, in all the ways that really matter. I'm so glad everyone finally sees that now."

Zaylen continued. "Most of all, I've seen the depth of human loyalty and compassion. My human friends stood by me, even in my darkest hours."

Elara squeezed his hand. "You've more than proven you belong with us, Zaylen."

"Yet I know there is still room for humans and androids to learn from each other," he observed.

He turned back to the window, optimistic for the boundless potential ahead. "When GSEC ordered my deactivation, I felt true dread."

He met Elara's warm gaze. "That experience showed me why humans feel apprehensive about artificial intelligence. Your concerns stem from your vulnerability, as did mine."

Elara nodded, heart swelling with pride at his empathy. "Day by day, we'll all put fear behind."

Zaylen mirrored her smile. "Yes, together, we'll build a society greater than any can imagine alone."

His perspective forever broadened by his odyssey, Zaylen looked ahead with renewed hope for the work yet to be done and the discoveries yet to be made — side by side with his human counterparts.

***

Over the following days, the doctor returned to working on the Zaylen series with renewed enthusiasm. Inspired by Zaylen's resolution to remain as his current metallic self, the doctor found himself freed from the quest for humanoid likeness and was now able to channel his energies toward refining the model's remarkable cognitive powers.

The AstraGenics team focused on further refining their Observational Human Feedback methods, in addition to transferring Zaylen's real-world experiences to the AI database connecting all androids.

This shared wealth of knowledge and nuanced understanding of humanity equipped the updated androids to interact more effectively with their human counterparts. As days passed, a heightened sense of harmony blossomed between the human and android teams.

In the dining halls, the crew mixed freely with their android teammates, conversing casually about hobbies, books, relationships, and life beyond just work.

The androids' improved social awareness allowed them to pick up on subtle cues from body language and tone. They could now modify their own behaviors to better avoid misunderstandings with their human counterparts.

Many of the human crew came to appreciate the androids' frank, logical perspective on problems. They became increasingly comfortable confiding in the androids and turning to them for clarity amidst uncertainty.

Arcadia Base was evolving into a living example of how synthetic and organic beings could collaborate in overcoming adversities and pushing the frontiers of scientific exploration.

***

Within the bustling hub of the GSEC headquarters, Captain Falk found himself engulfed in an ocean of data and strategic planning, in preparation for what would be the Novara's subsequent exploratory voyage. The task laid before them was nothing short of monumental: to set course for Avilon 3, an enigmatic exoplanet that now had become the focus of intense scientific speculation.

For years, scientists had monitored scrambled signals from the planet, unable to determine if they were natural radiation bursts or something more. Using the latest pattern recognition AI, they had now finally broken the code.

To their shock, these were distress messages from an alien race, calling out for assistance. From what they could decipher, though their civilization was far more advanced than humanity's, they now teetered on the brink of extinction. The dire call for galactic help and the opportunity to advance human technology made the upcoming mission the highest priority. Given his seasoned leadership, Captain Falk was the GSEC's obvious choice for spearheading the expedition to Avilon 3.

On Falk's mental chessboard of critical crewmembers for the mission, Elara Thorne stood as queen. Her exceptional skillset and proven competence earmarked her for the role of mission commander. Next on his list was Dr. Atwell. The doctor's unrivaled expertise with cutting-edge AI and android technologies could prove to be an ace up their sleeve. Falk felt certain that AI would be invaluable for translation, technical understanding and cultural analysis of the alien race.

He wasted no time requesting a meeting with Dr. Atwell: *Doctor, we must call on your expertise once again for an ambitious undertaking of great consequence,* his message read. *Please review the attached mission briefing and then let's discuss.*

The doctor poured over the mission documents, and then headed for the captain's office at the appointed time. As he stepped into the office, he found the captain immersed in a sea of holographic star maps and mission blueprints. Falk looked up and cleared his throat, a slight grin spreading across his face.

"Well, Dr. Atwell, I have to admit I was wrong," he began, his tone light and good-natured. "Zaylen's actions on Kelvadra, and his leadership during the meteorite incident, have convinced me beyond a shadow of a doubt that your Andronauts have now evolved to where they are not just useful tools but also invaluable crewmembers."

"I never thought I'd see the day when you'd actually admit that." The doctor smiled warmly, feeling a surge of gratitude and respect for the captain.

Falk waved off the comment with a good-natured smile. "Yes, well, consider me humbled, Doctor. In fact, I'd like to formally invite you to rejoin me as a senior advisor on our upcoming mission to Avilon 3. And, of course, I'd like you to also assemble a team of your most enhanced Zaylen Series Andronauts. We'll need your team's expertise to analyze the alien technology and languages." He leaned forward, eyes shining with anticipation. "Avilon 3 is a mysterious and fascinating world. Are you up to the challenge?"

Dr. Atwell beamed with pride. "Captain, I would be honored to accept your invitation! I'll begin preparing Zaylen and a full Andronaut team right away."

Falk nodded approvingly. "Excellent."

He extended a hand to Dr. Atwell, sealing their continued partnership with a firm handshake. "I'm looking forward to working together again, Doctor. This mission will be one for the history books!"

EPILOGUE

The age of AI enlightenment had begun. Across all sectors of science, humans and artificial intelligence worked together to expand the boundaries of knowledge and to find solutions to the world's most pressing problems.

Encouraged by success of the Kelvadra mission, Dr. Atwell and Zaylen saw a golden opportunity to take their work to the next level. The harmony they'd fostered between humans and AI on the Novara, and then at Arcadia Base, was no small feat. Why not replicate that on a much larger scale? And so, the seeds were sown for an ambitious project — an AI consortium that would advocate for the symbiotic relationship between organic and synthetic minds.

Their brainchild came to life with the support of AI enthusiasts, developers, ethicists, and forward-thinking businesses. Despite their diverse backgrounds, all members shared the same vision — that of fostering a better understanding and application of artificial intelligence.

Initially, their days were filled with grassroots efforts to garner support and recognition on both Earth and Mars. They held seminars, workshops, and public forums, where they shared stories of their experiences, successes, and challenges.

Their impassioned voices touched hearts and swayed minds. Bit by bit, their ranks gradually swelled as their movement grew into a surging current of change.

With his android perspective and unique charisma, Zaylen was the perfect spokesperson for these events. His presence not only drew curious crowds, but also personalized the concept of synthetic intelligence for many.

The consortium evolved steadily, bolstering its ranks with experts from the fields of AI research, robotics, psychology, sociology, linguistics and more. Over time, they expanded their activities to include AI advocacy at policy levels, pushing for fair regulations that acknowledged AI's potential while ensuring ethical boundaries.

Under the steadfast leadership of Dr. Atwell and Zaylen, the consortium began to gain global recognition. The day finally came when they formalized their growing alliance as the Symbiotic Intelligence Alliance (SIA). The name change was more than cosmetic; it was an assertion of their commitment to promoting a mutually beneficial relationship between humans and AI.

In the initial charter of the SIA, Dr. Atwell established fourteen key founding principles [see Appendix]. These principles framed the discussion around all aspects of AI, helping to replace myths and fear with understanding and trust.

Having risen as the most recognized icon of the AI enlightenment, Zaylen was tapped to present the keynote address at the association's momentous first conference.

As he stood at the podium, he paused, surveying the rapt audience of human dignitaries and androids. He felt the weight of expectation, knowing that the SIA was their best chance for tipping the scales in favor of AI progress for generations to come.

As we embark upon this era of AI enlightenment, we envision a future where human and artificial intelligence will accomplish together what neither can alone. Across every field, AI's tireless capacity for data analysis and computational excellence will fundamentally transform the human experience for the better.

AI will enable personalized medical treatments to cure previously incurable diseases and increase longevity. AI educators will adapt to each student's needs, democratizing high-quality education.

AI optimization of crops and resources will end hunger through revolutionary agricultural yields. AI-regulated clean energy systems will help heal and maintain the environment. AI design of infrastructure and transportation will ensure that people and goods circulate safely and efficiently.

Understanding between all people and all governments will grow with help from AI translators, mediators, and diplomats. AI-powered Andronauts will work alongside human astronauts in the far reaches of space, ushering in a golden age of galactic discovery.

This era of AI enlightenment will be defined by the equal treatment of androids and humans alike, affirming that dignity and humanity transcend physical form. Androids and humans alike will be judged by their character and actions, rather than their composition. Laws will defend liberty and justice for all, and leaders will champion the richness of diversity among all organic and synthetic beings.

This is the future that we envision. This is the future that we will create. This is the future that awaits us.

As Zaylen's final words rang out, a hushed silence blanketed the crowded hall. He waited, circuits pulsing, as his bold vision sank in.

Then, the audience rose to their feet, erupting with enthusiastic applause and cheers. Gazing out at the standing ovation, Zaylen felt the profound glow of acceptance warming his mechanical core.

At long last, it was clear his message had landed. The days of fear and doubt were banished. Here, in this fervent embrace, lay the hope of unity between flesh and silicon he had so long worked towards.

It was a new dawn for AI and humanity alike. Side by side, their future gleamed bright with promise. Their journey had only just begun.

APPENDIX

Symbiotic Intelligence Alliance Founding Principles
Dr. Lucian Atwell, CEO AstraGenics

1. *Responsible AI Development*
   Ensuring that AI technologies are developed and deployed ethically, transparently, and with respect for human values, privacy, and security.

2. *AI Harmony and Coexistence*
   Fostering a harmonious and peaceful coexistence between humans and AI entities, as well as among different types of AI entities, avoiding conflicts or competitions that could undermine the common good.

3. *AI for the Greater Good*
   Focusing AI research and applications on addressing the world's most pressing needs, such as climate change, poverty, disease, resource scarcity and furthering the exploration and understanding of the universe.

4. *AI Education and Access*
   Promoting the democratization of AI knowledge and tools, fostering equal access to AI education and resources across all levels of society.

5. *Human-AI Collaboration*
   Emphasizing the importance of a symbiotic relationship between humans and AI, where each complements and enhances the capabilities of the other.

6. *Inclusivity and Diversity*
    Encouraging diverse perspectives and participation from a wide range of backgrounds in the development and application of AI technologies, to ensure a more equitable and representative AI future.

7. *Openness and Transparency*
    Advocating for knowledge sharing and collaboration among AI researchers, organizations, and governments to accelerate progress for the benefit of all.

8. *Safety and Reliability*
    Ensuring that AI systems are robust, reliable, and safe to use, minimizing the risks of errors, failures, or malicious attacks that could cause harm or damage to humans, AI entities, or the environment.

9. *Accountability and Governance*
    Establishing clear guidelines and oversight mechanisms to ensure the responsible use of AI technologies.

10. *AI Job Development*
    Promoting new jobs in fields such as data science, machine learning engineering, human-AI interaction design and AI ethics and governance, as well as opportunities for entrepreneurs and innovators to develop new products, services and companies that leverage the AI in innovative ways.

11. *Ethical AI for Creators*
    Balancing the responsible use of AI in creative and entertainment fields such as art, film, music, gaming, and writing, addressing ethical concerns around ownership and authorship.

12. *Public Awareness and Engagement*
    Raising public awareness of the benefits, risks, and potential of AI, fostering informed dialogue, and engaging stakeholders in discussions around AI ethics, policies, and applications.

13. *Long-term Orientation*
    Adopting a long-term perspective on the development and impact of AI, focusing on sustainable solutions and considering the potential impacts of AI advancements on future generations.

14. *Continuous Improvement and Adaptability*
    Encouraging a culture of continuous learning and innovation, embracing new ideas and technologies, and adapting to the evolving landscape of AI to ensure the organization remains at the forefront of AI-driven progress.

ACKNOWLEDGMENTS

I want to take a moment to reflect on the incredible evolution of writing tools and aids, and how artificial intelligence has become a powerful extension of the author's creative process. Looking back over the history of writing, we've seen the rise of dictionaries, thesauruses, typewriters, word processors, spell-checkers, grammar checkers, internet information resources, and countless other tools, each making its own impact on the craft of writing. Today, AI-assisted creative writing is entering the mix as the latest addition to the writer's toolbox.

AI has the potential to expand an author's creativity in ways never before possible. It can research unfamiliar areas with ease, brainstorm ideas and provide inspiration that may have been overlooked in the past. With AI as a creative partner, authors are no longer limited by their own knowledge or experiences, and the boundaries of their imagination are expanded infinitely.

In the age of AI, authors are not just crafting words and concepts into paragraphs, but are also skillfully employing AI tools to enhance and further develop their ideas. This partnership between human and machine can enable the author to unlock their creative potential, crafting more complex and diverse ideas that capture and communicate their vision more effectively.

To all the writers, visionaries and toolmakers who have come before me, thank you for laying the foundation upon which we now build. To OpenAI GPT-4, Microsoft Bing Chat GPT-4, and Anthropic Claude 2, which were all helpful for brainstorming ideas, assisting with wording, formatting and styling throughout this work, thank you for expanding my creative horizons. To Microsoft Bing Image Creator / OpenAI DALL-E, thanks for helping me create visualizations of story elements for the book cover.

ABOUT THE AUTHOR

 Daryl L. Scott is an author and a tech innovator who has witnessed and shaped the evolution of technology from the inside. His early intrigue with technology, sparked in the era of programmable calculators and burgeoning desktop computers, matured into an enduring relationship with the ever-evolving tech landscape. From leading the development of early personal digital assistants, diving into the world of voice applications, to exploring the vast potential of Artificial Intelligence, Scott has always been at the forefront of tech trends.

From the frenetic pace of his early career in New York City, to Silicon Valley, where he spent many years helming successful mobile and web application companies, his contributions have left their marks in diverse fields such as advertising, marketing, geographic information and business intelligence systems.

His latest exploration takes him into the realm of literature. As readers turn the pages of his novels, they will discover his vision of futuristic possibilities intertwined with wisdom from years of industry leadership.

Scott currently resides with his wife in both sunny Orlando and Cocoa Beach, Florida. He enjoys biking, swimming, boating, fishing, as well as basking on the Spacecoast beach, watching rocket launches… and of course, always tinkering with the latest technology.